T0134898

Synthetic Data

Jimmy Nassif • Joe Tekli • Marc Kamradt

Synthetic Data

Revolutionizing the Industrial Metaverse

 Springer

Jimmy Nassif
Chief Technology Officer (CTO)
Idealworks GmbH
Munich, Germany

Joe Tekli
Associate Professor
Lebanese American University
Byblos, Lebanon

Marc Kamradt
Founder of SORDI.ai
Munich, Germany

ISBN 978-3-031-47559-7   ISBN 978-3-031-47560-3   (eBook)
https://doi.org/10.1007/978-3-031-47560-3

This Springer imprint is published by the registered company Springer Nature Switzerland AG
The registered company address is: Gewerbestrasse 11, 6330 Cham, Switzerland

Paper in this product is recyclable.

Foreword by Rev Lebaredian

For decades, computer scientists attempted and failed to develop computer vision algorithms that could reliably identify what is inside an image—telling us whether an image contains a cat or a dog. Most had given up hope that we would solve this problem in our lifetime. That was until the invention of AlexNet ignited the Big Bang of modern Artificial Intelligence. In 2012, Alex Krishevsky, Ilya Sutskever and Geoffrey Hinton had managed to take an idea that had its origin in the 1940s—Neural Networks—and apply it to the large amount of data available thanks to the Internet—the ImageNet database—with an extraordinary amount of compute readily available to any video gamer—NVIDIA's GPUs. AlexNet smashed all previous records in the ImageNet challenge, and within a few years, neural network-based algorithms evolved to achieve superhuman abilities in classification and computer vision. The method by which we develop the most advanced algorithms and software had fundamentally changed forever. Up until that moment, developing advanced software simply required an intelligent human, a small computer, a text editor and a compiler. AlexNet showed us that algorithms that were out of reach to humans were now possible. We could now write software that can write software—algorithms we are incapable of writing directly. The catch is that we need large amounts of the right data combined with enormous amounts of compute. The admission price into the AI game is data and compute.

Creatures such as humans are born into the world without a true understanding of their new surroundings. Human babies learn how to see and perceive the world through life experience. Babies learn how to perceive shapes, depth, color, sound, scents and taste. They learn how to identify their parents and siblings using all of their senses over a period of time. They also learn the rules of our world—otherwise known as physics—by conducting specialized experiments. Babies test the world by throwing glasses and utensils off the dinner table, breaking their toys and spilling liquids. They do this repeatedly until they develop an intuitive understanding of the rules.

AIs learn in precisely the same way. We feed them life experience—another way of saying **data**—during the training process. We teach them how to see, how to perceive and how best to manipulate the world around them by giving them millions of experiences. Unfortunately, it's impractical and in many cases unethical to have our AIs learn and gain these experiences in our world. We can't afford to allow our self-driving cars or industrial robots to learn how to drive and operate heavy machinery in the real world. It will take too long for them to gain the experience they need on the job; and in the process, they can cause too much harm as student drivers and heavy machine operators.

The solution to this problem is **simulation**. If we can construct digital worlds that are indistinguishable from the real world—worlds that look, sound, feel and behave exactly like our real world—we can generate an unlimited amount of life experience for our AIs. The more compute we throw at the simulation, the more life experience we can generate in the same amount of wall-clock time in the real world. AIs are free to learn without any risk of harm inside these simulations. They can learn to drive cars in simulations where they experience children running into the middle of the street, millions of times in varied lighting and weather conditions, without any harm coming to children in the real world. The data we generate in these simulations come with perfect labeling—labels that are impossible to gather accurately from the real world.

It turns out that the computing technology that sparked modern AI were originally and primarily designed for simulating virtual worlds. This of course is the programmable GPUs initially developed for powering 3D computer graphics and rendering for interactive video games. Modern video games are in essence, simulations of fantastic virtual worlds. The most advanced video games approach the real world in complexity and physical accuracy.

There's a beautiful duality to 3D computer graphics and computer vision. 3D computer graphics is a function that transforms a structured description of a 3D world into images over time—a simulation of what a camera sensor would experience in that world. Computer vision is the inverse of this function; transforming images over time into a structured representation of the 3D world. The AIs that will be the most impactful and valuable to human-kind will be the ones that can understand our real world and operate within it. To create these AIs, we must first model the real world and simulate it to generate the life experience for them. Once our AIs achieve proficiency in understanding and manipulating worlds, they will then assist us in designing efficient, sustainable and delightful virtual worlds—worlds that will act as the blueprints for what we choose to build in our real world.

Inevitably, industries will use these virtual worlds to optimize their plans, designs, factory floors, robotics systems and logistics operations; long before production starts in the physical world. This is at the heart of the NVIDIA Omniverse platform for building and operating industrial metaverse simulations. Cloud computing, AI and simulation technologies are converging, giving industries the superpower of planning and optimizing their factories in digital worlds—turning real-world problems into software problems. Manufacturing companies will use these superpowers to increase throughput, maximize quality, optimizing resource consumption and greatly reduce time-to-market—all while achieving challenging sustainability goals.

This book has arrived at just the right time. Every industry now understands that AI will fundamentally change how we design and build everything, but very few understand what is needed to create the specialized AIs. This book examines how multimedia data and digital images in particular are inputs into the creation of fully virtualized worlds in the form of digital twin factories and fully digitalized industrial assets. It relies on practical use cases from the automotive and manufacturing industries and their digitalization technologies based on the SORDI dataset. With this book, you will understand the nature of data and its unique value for AI. You will learn how to capture, structure and generate data essential for AI in order to build the industrial metaverse.

Jimmy Nassif, Chief Technology Officer at Idealworks and co-Founder of SORDI.ai, is a key contributor to the successful adoption of NVIDIA Omniverse at BMW Group. In this book, Jimmy provides his unique experience and insights on how to leverage synthetic data to build the industrial digital twins of tomorrow. Joe Tekli, Computer Engineering Professor at the Lebanese American University, has supported the BMW Group for many years as external lead collaborator. He provides his academic perspective, adding breadth and depth in describing state-of-the-art digitalization technologies and building the SORDI dataset. Marc Kamradt, Head of BMW Group TechOffice and cofounder of SORDI.ai, completes the book with a look at the ongoing and future digitization activities enabled by the generation and use of synthetic data, and the trends driving the industrial world today.

Rev Lebaredian
Vice President
Omniverse & Simulation Technology at NVIDIA
Santa Clara, CA, USA

Foreword by Dr. Dirk Dreher

Industrial production networks involve high levels of industrialization experience and skills in terms of product knowledge and product integration. State-of-the-art production facilities are geared toward lean manufacturing, sustainability, and digitalization. Cutting-edge industrial processes aim at maximizing levels of digitalization, flexibility, and automation, in order to serving fast-changing markets and customer needs, while simultaneously minimizing resource consumption. On a journey of a transformation to adapt to future challenges, there is a need for a culture of continuous improvement and agility to shape a "lean, green and digital" production. These concepts are at the core of BMW Group's "iFactory". "Lean" aims at achieving high efficiency, precision, and flexibility. "Green" aims at systemizing production with minimum resource usage. "Digital" aims at using cutting edge digitalization technologies including artificial intelligence (AI), big data processing, and virtualization, to improve planning, execution, problem solving, and product quality.

According to its Next Level Mobility report for 2022, BMW Group's iFactory aims at digitalizing not only the next-generation BMW automobiles, but more importantly the company's internal processes and operations, pioneering the next-generation digital company. The creation of digital worlds, in the form of Digital Twins simulated in the metaverse, harnessing billions of data points and 3D visualizations of a factory provides an intuitive way into production data. A suite of digital tools, connecting to this Digital Twin, will drive a digital transformation and a data-driven and learning organization. From the virtualization of the internal structures and processes of a factory, to virtualizing the manufactured products, the benefits of digitalization span all industrial stakeholders. Providing transparency to the management and delivering virtual design, monitoring, tracking, and forecast capabilities for engineers, the digital worlds of the metaverse are transforming the way we are manufacturing in the twenty-first century. Multiple data sources are offered from a Digital Twin to the work organization for fast problem solving and to reach operational excellence. Furthermore, digitalization affords high flexibility and customization for clients.

Nonetheless, yoking the power of digitalization goes beyond individual digitalization elements such as vision cameras, the deployment of a computer vision model infrastructure, the purchase of Cloud resources, or the development of a virtual simulation of the factory floor. Industries with a thorough digitalization strategy focus on developing digital skills at all levels of a company on the one hand. On the other hand, digital champions are taking a holistic, data-centric approach that uses real and increasingly synthetic data. In particular, the generation and use of synthetic data opens up new opportunities in numerous areas and will play a key role in the development of AI applications.

This book addresses industrial digitalization and the usage of state-of-the-art digitalization technologies to create full-fledged virtual factories of the future. It describes BMW Group's Synthetic Object Recognition Dataset for Industry (SORDI), as one of the main catalysts for creating the next-generation "robot gym": training robots in the virtual world to prepare them to execute in the physical world. The first section provides a background of the origins and evolution of record keeping, and the introduction of industrial multimedia data. It describes the transition from simple digitization of data toward a full digital transformation of industrial processes. The following section introduces the concept of a Digital Twin and how to make it work to benefit a modern factory. It describes how the Digital Twin is controlled, automated, and how the generation of synthetic data to be used for training automated robotic systems is created. The Digital Twin is viewed as a "robot gym", training virtual robotic twins using synthetic data in order to allow their real-world counterparts to work and execute their tasks in the real world. Different challenges and responsibilities that arise with the introduction and usage of new digitalization technologies and their impact on industry are also addressed, including changes in jobs, wages, talents, and the carbon footprint.

This book offers a fresh and modern look at today's digital industry, specifically focusing on using artificial data in smart manufacturing and the industrial metaverse. It's authored by three individuals with extensive knowledge in the field. Jimmy Nassif, who is the Chief Technology Officer at Idealworks and co-founder of SORDI.ai, played a key role in incorporating NVIDIA Omniverse at BMW Group. He shares his exceptional expertise and insights on utilizing artificial data to create advanced digital versions of industrial elements. Joining him is Joe Tekli, Computer Engineering Professor at the Lebanese American University, who has been a long-standing collaborator for BMW Group. Tekli brings an academic perspective that dives into cutting-edge digitalization technologies and the development of the SORDI dataset. Furthermore, Marc Kamradt, the Head of BMW Group TechOffice and co-founder of SORDI.ai, explores future digitalization efforts driven by the generation and application of artificial data. He also provides valuable insight into ongoing digitization efforts that leverage AI capabilities such as creating realistic virtual worlds through tools like Omniverse, while also addressing current trends, shaping modern-day businesses worldwide.

This manuscript is undeniably outstanding and poised to serve as a concise guide for innovative pioneers of the twenty-first-century tech industry.

Dirk Dreher
Chief Executive Officer
BMW Hams Hall Motoren GmbH
Coleshill, UK

Preface

The industrial world is at the verge of a new manufacturing movement – that of full-fledged digitalization and the industrial metaverse. Digitalization is one of the top challenges of modern industries of this decade. From AI and IoT, to Mixed Reality, Digital Twins, and ultimately the Metaverse, the integration and smart usage of digitalization technologies is bound to fuel the growth of industry and the optimization of its processes. During the last iteration of NVIDIA's GTC conference in March 2023, BMW Group announced the expansion of its usage of NVIDIA Omniverse across its production network around the world, officially opening the automaker's first entirely virtual factory powered by Omniverse. Omniverse is the zenith of more than 25 years of NVIDIA graphics, computing, AI, and simulation technologies, which allows industries to plan and optimize their manufacturing projects entirely virtually. We are talking about virtualizing the whole thing: from the smallest bolt and cutter, to the largest conveyor belt and assembly floor, all the way through the manufacturing and logistics robots and operators in between. This allows manufacturing companies to reach production quicker, while optimizing resource consumption, improving time-to-market, and enabling improved sustainability.

These are few of the main ideas and concepts described in this book, which focuses on the impact of digitalization and digital transformation technologies on the Industry 4.0 and smart factories, how the factory of tomorrow can be designed, built, and run virtually as a digital twin likeness of its real-world counterpart, before the physical structure is actually erected. It highlights the main digitalization technologies that have stimulated the Industry 4.0, how these technologies work and integrate with each other, and how they are shaping the industry of the future. It describes how digital images are used to create fully virtualized worlds in the form of digital twin factories and fully virtualized industrial layouts. It uses BMW Group's latest SORDI dataset (Synthetic Object Recognition Dataset for Industry), i.e., the largest industrial images dataset to-date and its applications at BMW Group, as one of the main explanatory scenarios throughout the book. It also emphasizes the need for synthetic data to train advanced deep learning computer vision models, and how such datasets will help create the "robot gym" of the future: training robots on synthetic images to prepare them to function in the real world.

Writing this book would not have been possible without the hard work of the research and development teams at IdealWorks and BMW Group TechOffice, with the support of the Lebanese American University's School of Engineering, whose engineers, developers, and graduate/senior students have been successfully collaborating for many years on the development of many of the industrial digitalization solutions mentioned in the book. We are also profoundly grateful to Rev Lebaredian, Vice President, Omniverse & Simulation Technology at NVIDIA, USA, and Dr. Dirk Dreher, Managing Director BMW Hams Hall Motoren GmbH at BMW Group, Germany, for agreeing to review our book and for endorsing it through their eloquent and powerful forewords. We are sincerely grateful and extremely proud to have their support. We also like to extend our deepest gratitude to the experts and colleagues who volunteered to review our book, providing us with constructive comments on how to improve its content and organization: Pr. Raphael Couturier, Ph.D., Head of the AND team, University of Franche-Comté, Belfort, France; Jibran Jahshan, vice President, Software, NVIDIA Corporation, USA; Dr. Christian Imgrund, Ph.D., Senior Advisor and former BMW Manager for planning and production, Munich, Germany; Pr. Yannis Manolopoulos, Professor in the Department of Informatics, Aristotle University of Thessaloniki, Data Science & Engineering Laboratory, Thessaloniki, Greece; Marco Prueglmeier, Founder Noyes Technologies, Senior Logistics Advisor and former BMW Executive Manager for logistics, Munich, Germany; Dr. Bechara Al Bouna, Ph.D., Owner and Chief Executive Officer of InMind.ai, Beirut, Lebanon; and Dr. Boulos Al Asmar, Ph.D., Senior Vice President of Engineering, Idealworks GmbH, Munich, Germany.

The authors believe this book will benefit industry stakeholders and managers, and technology enthusiasts, providing them with a comprehensive, well structured, and easy-to-read introduction and description of the world and future of industrial digitalization. The authors' collective own opinion is reflected throughout the content presented in this book. We hope that the unified presentation of industrial digitalization in this book will contribute to strengthen further development and research on the subject matter.

Munich, Germany	Jimmy Nassif
Byblos, Lebanon	Joe Tekli
Munich, Germany	Marc Kamradt

Keywords

Industrial metaverse; SORDI; Industrial data; Synthetic data; Digitalization; Digital transformation; Smart Manufacturing; Industry 4.0; Industry 5.0; Robot gym

Abstract

- The book concentrates on the impact of digitalization and digital transformation technologies on the Industry 4.0 and smart factories, how the factory of tomorrow can be designed, built, and run virtually as a digital twin likeness of its real-world counterpart, before the physical structure is actually erected.
- It highlights the main digitalization technologies that have stimulated the Industry 4.0, how these technologies work and integrate with each other, and how they are shaping the industry of the future.
- It examines how multimedia data and digital images in particular are being leveraged to create fully virtualized worlds in the form of digital twin factories and fully virtualized industrial assets. It uses BMW Group's latest SORDI dataset (Synthetic Object Recognition Dataset for Industry), i.e., the largest industrial images dataset to-date and its applications at BMW Group, as one of the main explanatory scenarios throughout the book.
- It discusses the need of synthetic data to train advanced deep learning computer vision models, and how such datasets will help create the "robot gym" of the future: training robots on synthetic images to prepare them to function in the real world.

Contents

About the Authors

Jimmy Nassif is currently serving as Chief Technology Officer (CTO) at Idealworks GmbH (since 2020), a wholly owned subsidiary of BMW Group. The mission of Idealworks is to develop solutions for autonomous logistics, building intelligent, flexible, and collaborative autonomous mobile robots. From hardware to software, to the smart factory cloud platform, the activities of Idealworks are validated in BMW Group production environments, producing solutions which are centered on the customers' needs to reliably improve safety and efficiency across their facilities.

J. Nassif is co-founder of SORDI.ai, Synthetic Object Recognition Dataset for Industries. He has more than 15 years of experience in automotive engineering. He previously served as Head of Information Technology (IT) Planning and Product Owner Logistics Planning at BMW Group (2018-2022). He also served as Innovation Manager and Head of Logistics Virtual Reality (VR), Mixed Reality (VR/MR), Cognitive Computing (CC), and Artificial Intelligence (AI) within BMW Group's Logistics division (2014-2020). His activities revolve around leveraging and integrating innovative solutions in the above mention fields to solve real-world problems in logistics, distribution, and manufacturing. J. Nassif obtained his M.Sc. degree in Mechatronics, Robotics, and Automation from Technical University of Munich (TUM, 2007). He pursued his Ph.D. as part of the BMW Ph.D. program from 2008 to 2011. His recent research activities focus on understanding the impact of digitalization and simulation on logistics planning, and finding concrete value, pushing towards accelerated rollout and development cycles within the industry.

Joe Tekli is an Associate Professor of Computer Engineering in the Lebanese American University (LAU). He was appointed as Assistant Provost for Strategic Planning and Academic Initiatives and Partnerships in September 2023. He obtained his M.Sc. and Ph.D. degrees from the University of Bourgogne (UB), LE2I CNRS, France (2009), both awarded with Highest Honors. He has completed various post-docs and visiting scholar research missions: University of Michigan (UMich), USA (Summer 2018); University of Sao Paulo (USP- ICMC), Brazil (2010-2011); University of Shizuoka, Japan (Spring 2010); and University of Milan, Italy (Fall 2009). He was awarded various prestigious fellowships: Fulbright (USA), FAPESP (Brazil), JSPS (Japan), Fondazione Cariplo (Italy), French Ministry of Education (France), and Association of Francophone Universities (AUF, Canada). His research covers semi-structured, semantic, and multimedia data processing, data-mining, and information retrieval. He has coordinated various international research projects and has more than 65 publications in prestigious journals and conferences.

J. Tekli has initiated a collaboration between LAU and BMW Group (since 2018), offering engineering students internships at the company's seat in Munich, and promoting collaborative projects between both institutions. He co-founded and is currently serving as Director of the InMind Academy (since 2022), a professional training program in collaboration with BMW Group, Idealworks, and InMind.ai. He is also a collaborator on SORDI.ai, Synthetic Object Recognition Dataset for Industries.

J. Tekli is a senior member of IEEE, a member of ACM, and is currently serving as the Vice-Chair of ACM SIGAPP French chapter (since 2018), Associate Editor of Springer KAIS journal, and member of the US-Atlantic Council AI Connect initiative (since 2022).

Marc Kamradt is currently serving as Head of BMW Group TechOffice, Munich (since 2021). The work of the TechOffice mainly revolves around the development and integration of digitalization technologies to help build next-generation smart factories, including Ominverse Artificial Intelligence (AI) pipeline, synthetic data generation, and BMW Green physics AI.

M. Kamradt is founder of SORDI.ai, Synthetic Object Recognition Dataset for Industries. He was previously appointed to different leadership positions within BMW Group. He served as Senior Expert Innovation, Lead BMW Innovation Lab (2016-2020), leading projects around visual object recognition, knowledge graph enabled manufacturing assistance, and chat bot production at BMW Group. He also served as Innovations Manager (2013-2016), leading projects around cognitive production control and error analysis, and knowledge graph enabled manufacturing assistance. Prior to his leadership roles, he had an extensive technical experience as Enterprise Information Technology (IT) Architect at BMW Group (2011-2012), Product and Process Planning Electronics at BMW Group (2007-2010), IT Specialist at BMW Group (2002-2007), and IT Specialist at T-Systems International GmbH (2000-2002).

M. Kamradt obtained his Computer Science diploma from Carl von Ossietzky University of Oldenburg, Germany (1999).

Abbreviations

AI	Artificial Intelligence
AMR	Automatic Mobile Robots
ANN	Artificial Neural Network
API	Application Programming Interface
AR	Augmented Reality
ASCII	American Standard Code for Information Interchange
AWS	Amazon Web Services
BAR	Bounding-box Automated Refinement
BJT	Bipolar Junction Transistor
CAD	Computer-Aided Design
CB	Contextual Bandits
CBDM	Cloud-Based Design Manufacturing
CBOW	Contiguous Bag Of Words
CCR	Constrained Capture Randomization
CEO	Chief Executive Officer
CIMS	Computer-Integrated Manufacturing Systems
CNN	Convolutional Neural Network
CODASYL	Conference/Committee on Data Systems Languages
CTM	Color Texture Moments
CTO	Chief Technology Officer
CV	Computer Vision
CPS	Cyber Physical System
DAC	Design Automated by Computer
DB	DataBase
DL	Deep Learning
DP-RNN	Dual Path Recurrent Neural Network
ECM	Enterprise Content Management
ENIAC	Electronic Numerical Integrator and Computer
EPC	Electronic Product Code
FCR	Full Capture Randomization
GAN	Generative Adversarial Network

GPT	Generative Pre-trained Transformer
GPU	Graphical Processing Unit
HTML	Hypertext Markup Language
IaaS	Infrastructure as a Service
IC	Integrated Circuit
IDC	International Data Corporation
IMS	Information Management System
IoT	Internet of Things
IoU	Intersection-over-Union
IP/TCP	Internet Protocol/Transmission Control Protocol
IPv6	Internet Protocol version 6
IR	Information Retrieval
ISP	Internet Service Provider
IT	Information Technology
JSON	JavaScript Object Notation
KB	Knowledge Base
KLT	Kleinladungstrager small load carrier box
LoA	Level of Accuracy
LoD	Level of Development
LoR	Level of Recognizability
LSA	Latent Semantic Analysis
LSH	Locality Sensitive Hashing
ML	Machine Learning
MLI	Mixed-Lighting Illumination
MOS	Metal Oxide Semi-conductor
MPEG-7	Moving Picture Experts Group – Multimedia Content Description Interface
MR	Mixed Reality
NASA	National Aeronautics and Space Agency
NFT	Natural Feature Tracking
NIST	National Institute of Standards and Technology
NLP	Natural Language Processing
NoSQL	Not Only SQL
NSF	National Science Foundation
ORM	Object Relation Modelling
OT	Operation Technology
OWL	Ontology Markup Language
PaaS	Platform as a Service
PCB	Printed Circuit Board
PLC	Programmable Logic Controllers
PSNR	Peak Signal to Noise Ratio
R&R	Restoration and Recognition
RDF/S	Resource Description Framework / Schema
RFID	Radio-Frequency Identification
RGB-D	Red Green Blue – Depth

RL	Reinforcement Learning
RNN	Recurrent Neural Network
RoI	Region of Interest
ROS	Robot Operating System
SaaS	Software as a Service
SC	Static Capture
SCADA	Supervisory Control And Data Acquisition
SDK	Software Development Kit
SDMP	Sequential Decision Making Problem
SDR	Structured Domain Randomization
SEAC	Standards Eastern Automatic Computer
SeC	Sequential Capture
SLAM	Simultaneous Localization And Mapping
SNM	Structure Natural Measure
SOA	Service-Oriented Architectures
SORDI	Synthetic Object Recognition Dataset for Industry
SPARQL	SPARQL Protocol and RDF Query Language
SQL	Structured Query Language
SSD	Solid State Drive
SSIM	Structural Similarity Index Measure
SW	Semantic Web
TMS	Toyota Production System
Transfer Learning	Transfer Learning
TSDB	Time Series DataBase
URL	Uniform Resource Locator
USAF	United States Air Force
USD	Universal Scene Descriptor
VAE	Variational AutoEncoders
VFL	Virtual Factory Layouts
VR	Virtual Reality
WNA	WordNet Affect
XML	Extensible Markup Language

Chapter 1
Welcome to the Age of Data

For the past few decades, data has become increasingly available in digital format, especially on the Web, considered as the largest multimedia database to date. Its applications include video-on-demand systems, video conferencing, medical imaging, on-line encyclopedias, cartography, image retrieval, among others. State of the art storage, indexing, and retrieval capabilities have made it possible to efficiently process huge amounts of data, and have transformed data into the most valuable corporate resource – the new oil of the twenty-first century [12]. With the exception of Saudi Aramco ranking at #1, the four tech titans – Apple, Microsoft, Alphabet (Google and YouTube's parent company), and Amazon remain unrivalled as the top most valuable companies in the world in 2022 [35]. Their profits are rising, collectively recording staggering revenues accumulating over $264.7bn in net profit in the last quarter of 2021 [35]. Few of us can not survive without Google's search engine, Microsoft's office tools, Amazon's one-day delivery, or YouTube's videos. What is even more fascinating is that many of these firms provide most of their services for free, or so it seems: users pay in effect by sharing more data. According to the former CEO of Google Eric Schmidt, humankind generated 5 exabytes of data from the dawn of civilization until 2003 [45]. Today, our best approximations suggest that around 2.5 quintillion bytes of data are produced every day – that's 2.5 followed by an astounding 18 zeros [41]. Undoubtedly, there are genuine concerns about how these tech giants are using our data and whether they are exploiting what they know about us. Yet there is also no denying of the potential and the positive impact that data is having on the world, from improved healthcare, to creating new jobs, creating online communities, cutting back on pollution and energy waste, to automotive vehicles, smart homes, and smart city management.

Big data has also been a prime driver of smart industries with many new opportunities in logistics, manufacturing, and production. Data from various industrial equipment and manufacturing devices equipped with dedicated sensors and

J. Nassif et al., *Synthetic Data*, https://doi.org/10.1007/978-3-031-47560-3_1

controllers, is regularly generated and ingested into smart industrial systems to provide real-time response from the physical world, such as dynamic supply chain scheduling [38], production plant management, predictive maintenance [5], as well as diagnosis, prognosis and anomaly detection in the industrial pipeline [53, 54]. Another major source of industrial data is databases and data repositories including data logs and historical records of the industry's past and ongoing operations. Compared with the dynamic data streamed from sensor-enabled devices which allow monitoring the current status of an industrial process, database records are processed offline to generate detailed analytics that allow long-term performance prediction and enhancement in industrial operations such as improved project planning [50], manufacturing network design [18], and critical anomaly prediction [23]. Both historical data and real-time streaming data can be integrated to train AI (artificial intelligence) models to generate useful insights, predict future events, and simulate real-world industrial scenarios in virtual environments. Using state of the art digital transformation technologies, the smart factory of tomorrow can be designed, built, and run virtually as a digital twin likeness of its real-world counterpart, before the physical structure is actually erected. This will allow the factory managers and decision makers to gain more insights on the functioning of the factory and its complex and interrelated systems, in order to make better decisions when building the actual physical structure.

***

But how did we reach the age of data? How did we go from stone tablets, paper scrolls, and the Epic of Gilgamesh, to silicone chips, the cloud, and the Internet of Things? How can we handle multimedia industrial data? And what are the main prospects and the key challenges of big data in building smart industries?

***

We attempt to answer these questions in this chapter and in the remainder of this book...

1.1 Origins and Evolution of Record Keeping

The Pyramid Texts, dating back to 2400–2300 BC, are regarded as the oldest known ancient texts, consisting of funerary religious transcripts carved onto the subterranean stone walls of the Egyptian Pyramids at Saqqara on the Giza plateau [24]. The Epic of Galgamesh, written on clay tablets somewhere between 2100–1200 BC, is regarded as the earliest surviving piece of literature, consisting of an epic poem from ancient Mesopotamia. The Code of Hammurabi in Babylon (1792 BC),

Table 1.1 Sample legal cases covered in the Code of Hammurabi [44]

Legal area	Codes of law
Offences against the administration of law	If any one ensnare another, putting a ban upon him, but he can not prove it, then he that ensnared him shall be put to death (#1)
	If any one bring an accusation of any crime before the elders, and does not prove what he has charged, he shall, if it be a capital offense charged, be put to death (#3)
Property offenses	If any one break a hole into a house (break in to steal), he shall be put to death before that hole and be buried (# 21)
	22. If any one is committing a robbery and is caught, then he shall be put to death (#22)
Commerce	If a merchant entrust money to an agent (broker) for some investment, and the broker suffer a loss in the place to which he goes, he shall make good the capital to the merchant (# 102)
	If a merchant gives an agent corn, wool, oil, or any other goods to transport, the agent shall give a receipt for the amount, and compensate the merchant therefor. Then he shall obtain a receipt form the merchant for the money that he gives the merchant (#104)
Assault	If a man puts out the eye of another man, his eye shall be put out. [an eye for an eye] (#196)
	If he breaks another man's bone, his bone shall be broken (#197)
Professional men	If a builder builds a house for some one, even though he has not yet completed it; if then the walls seem toppling, the builder must make the walls solid from his own means (#233)
	If a sailor wreck any one's ship, but saves it, he shall pay the half of its value in money (238)

written on Basalt rock, includes statements that govern the handling and keeping of records and property (cf. Table 1.1). Business and commercial records have also been recorded as far back as 3000 BC in ancient Sumeria, where receipts in the form of clay tablets that were exchanged and then stored for record keeping [31]. Babylonian loan records have been found from the eighteenth century BC [34]. Similar records were also recovered in ancient Assyria, India, and China dating back to around 2000–1500 BC. Later, in ancient Greece and during the Roman Empire, loan records were stored on papyrus and paper scrolls and stored in temples. Tally sticks were later introduced in medieval Europe, where a stick was marked with notches and then split lengthways. The two halves bear the same notches and each party to the transaction received one half of the stick as proof. The sticks were then stored for future reference and claims. The tally stick was used by the British government for managing taxes, until the early nineteenth century, where the last tally stick stores were ordered for destruction by burning, and were replaced with paper [31]. Paper, and in earlier times papyrus and vellum, were progressively used through the centuries to record all sorts of data, from religious texts and literature, to transactions, contracts, bills, and business deals. Often, the records were signed and sometimes sealed in wax with the marks of the stakeholders involved. As computers came into commercial use in the twentieth century, businesses began to computerize their systems – which required the conversion of the real-world paper records into a representation that computers could understand.

1.1.1 The Advent of Digital Computers

Since the 1950s when digital data came to being, computer systems and computing technologies and their applications have evolved drastically. In 1950, the first generation electronic computer SEAC[1] was created in the USA [43]. Compared with its revolutionary predecessors like ENIAC[2] and the German Z3,[3] SEAC was arguably the first fully operation digital computer that adopted the, now ubiquitous, von Neumann architecture scheme (cf. Fig. 1.1). That same year, Russell Kirsch, an American engineer at the National Bureau of Standards, used a rotating drum scanner and photomultiplier connected to SEAC to create the first digital image from a photo of his infant son (cf. Fig. 1.2) [2, 13]. The image was stored in the SEAC memory by an electronic staticizer and was viewed via a cathode ray oscilloscope [26]. The ethereal black-and-white photo only measured 176x176

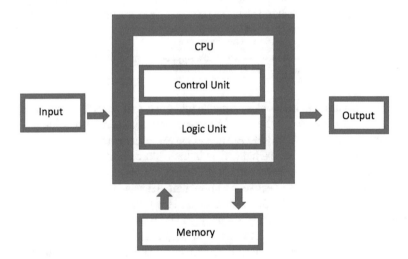

Fig. 1.1 The von Neumann computer architecture scheme, widely adopted in most computer systems since the 1950s [14]

---

[1] Standards Eastern Automatic Computer.

[2] *Electronic Numerical Integrator and Computer* was the first programmable, electronic, and general-purpose digital computer, completed in 1945. Dedicated at the University of Pennsylvania and accepted by U.S. Army Ordnance Corps in 1946, its first program was a study of the feasibility of the thermonuclear weapon.

[3] Z3 was a German electromechanical computer designed by German civil engineer and pioneering computer scientist Konrad Zuse in 1938. Completed in 1941, it was the world's first working programmable and fully automatic computer.

Fig. 1.2 The first digitally
scanned image of Russel
Kirch's three-month-old
son Walden, 1957 [40]

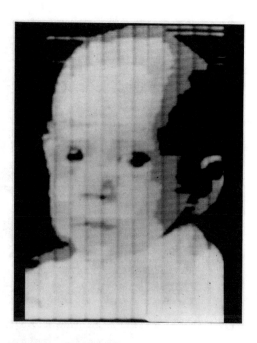

Fig. 1.3 Data entry in the 1950s [1]

pixels – compared with today's multicolored megapixel digital photos. But it became the point of origin for all computer imaging to follow, and it was dubbed in 2003 by Life magazine as one of "the 100 photographs that changed the world". Despite the early breakthroughs of the 1950s, digital storage space remained extremely limited well into the 1960s, where business information was still mostly on paper. People and input-output devices like punched cards were needed to translate data from paper into a digital format that computers can process (cf. Fig. 1.3).

Fig. 1.4 Shipping a 5 MB IBM disk drive in 1956 [22]

Data records were stored on paper punched cards, which were also used for input (scanning the holes in the cards) and output (punching holes in blank cards). Human operators typed the content of the paper records onto cards, so the computer could read and consume the information. In the late 1960s, magnetic tape and consequently disk drive storage gradually replaced punched cards in large computer systems. With the advent of disk storage (Fig. 1.4), the ability to access data directly and speedily became a possibility, as individual portions of a disk are addressable programmatically. Prior to the existence of disks, most data processing took place in batches where data was processed sequentially following the order it was stored in on punched cards or on magnetic tape. Disk drive technology enabled random data access.

1.1.2   The Advent of Corporate Database Systems

In the late 1960s, early database systems were developed to manage data stored on disk that could be randomly accessed and updated. Prior to databases, data was stored and managed in files which could only be accessed sequentially.[4] Two of the most common database structures used were the network model (e.g., CODASYL)

---

[4] File systems are still used today in certain applications, such as managing data in operating systems (e.g., New Technology File System/NTFS with Microsoft Windows, and Apple File System/ AFS with MacOS).

and the hierarchical model (e.g., IMS) [31]. Before storing data in the database, a data design phase was performed by data administrators to transform the business data, still on paper in that era, into hierarchical or network models. In the 1970s, the vastly popular relational database model was introduced [7], which continues to dominate business systems in the twenty-first century. Yet data was still mostly paper based in the 1970s and 1980s and had to be transformed, often by scanners or operators re-typing forms, to be stored in databases. This meant the data had to be re-structured and re-organized following the principles of database systems. Also, databases usually stored any piece of information exactly once – the latest version only, making it difficult to perform audits and historical analyses. This was mostly emphasized in the so-called *relational* (or SQL) database model, which remains the most widely used legacy database model used by many big corporations to-date (e.g., industries, banks, universities, etc.). For example, a *sales receipt* would be deconstructed into its constituent pieces including: *client data, supplier data, purchase order, purchased items*, etc. where each piece is stored in its own structure (or so-called table). All constituent tables would be connected together using well defined joins to acquire the original *sales receipt* data. This ensures that multiple receipts by the same client from the same supplier purchasing the same items would only hold the client, supplier, and item data only once, without needless repetitions and duplications. Putting things into perspective, there was a dire need to save in storage space due to the huge prices of disk storage back then. At the time, a popular disk storage device was the 3330 model 11 which stored 200 MBs and whose price ranged from $74,000 to $87,000 (in 1970s USD) [42]. In other words, 1 MB of disk storage cost around $160 (in 1970s USD), the equivalent of thousands of dollars in 2022 [31]. Yet this would prove extremely limiting later on with the rise of artificial intelligence techniques which rely on the historical data for training and prediction. Another issue with database solutions is the lack of security information associated with stakeholder signature and authentication seals. For example, storing a collection of receipts in a database required storing the data itself in a database system, and scanning the original paper receipts, and storing their signatures or seals in a separate document system for authentication purposes. Modern business processes were also increasingly requiring the data to be kept for a certain number of years, especially for claims in the case of disputes. Hence, a new category of software systems called Enterprise Content Management (ECM) was developed in the early 1990s to store digital images of paper records. In other words, if an exact copy of the real-world paper document was needed, separate database and ECM systems were put in place to do the job, causing the same data to be stored more than once.

1.1.3   From Data Warehousing to Big Data

As processing power and storage memory increased in the late 1980s and 1990s, coupled with a sharp decrease in cost, companies could afford to gather and analyze large amounts of historical business data, such as sales, manufacturing, and employment records. Data warehouses were introduced as a special type of data-bases with simple data representations designed for intuitive and high-performance retrieval [25]. The initial database paradigm of having well-structured data repre-sentations in order to minimize storage space started slowly shifting toward more loosely coupled data representations. So-called "denormalized" schemas become more and more popular for data warehousing. A denormalized schema allows the data to be represented as a whole, similarly to its original representation, versus breaking it down into smaller pieces the way legacy databases usually do it [48, 49]. For example, a *sales receipt* would be stored as one big data record, instead of breaking it down into *client data, supplier data, purchase order, purchased items*, etc. All the latter fields would be acceded through the receipt itself, where the client, supplier, and item data would be duplicated and repeated in every receipt. This would allow different branches or departments of a company, or different compa-nies who produce different kinds of receipts, to easily aggregate and mine all their *sales receipts* data together, since every *sales receipt* is self-contained and complete regardless of its constituent items and inner structure. As a result, the late 1990s and 2000s saw more and more legacy database systems connected with each other and distributed over multiple computer system and networks, using denormalized and loosely structured data warehouse architectures. In the 2010s, loosely structured and denormalized databases – so-called NoSQL (referring to the well-structured legacy SQL databases), saw increased usage, implemented following parallel, sparse, distributed, and multi-dimensional architectures. This new generation of databases (like MongoDB, Apache HBase, and Appache Cassandra) were designed for scalability to very large data volumes and for distribution over hundreds of thou-sands of computer systems. Their main intention is optimization for efficient and scalable data access, such that a single read operation can retrieve all fields that belong to a logical business record. For instance, the column NoSQL database Cassandra is used to store event data of automation controller [17], the document NoSQL database MongoDB is used to store machine data [47], and time-series databases (TSDB) are receiving growing attention in handling sensor data [9]. Nowadays, loosely structured NoSQL databases, coupled with unlimited cloud stor-age and computing resources, are paving the way to the handling of big data and real-time multimedia Web applications.

1.2 Handling Multimedia Industrial Data

The realization of the Internet of Things (IoT) vision of collaborative cyber-physical systems, where physical machines and software agents meaningfully and intelli-gently manipulate and exchange information and services without human interac-tion, remains in its early stages. Nonetheless, it is currently unfolding in the industrial world, especially in *smart manufacturing* which is defined by NIST[5] as a

---

[5] National Institute of Standards and Technology.

completely integrated, collaborative manufacturing digital ecosystem that responds in real-time to meet changing demands and conditions in the factory, in the supply network and in customer needs [29, 39]. Yet a major problem facing autonomous and collaborative data processing in smart manufacturing and industrial applications is the nature of shared multimedia data sources, which often exist in loosely distributed environments, with unstructured and heterogeneous contents, created by different users (e.g., terminals, sensors, and agents), developed by different vendors, with different profiles, formatted following different standards, and using different interfaces or protocols. Add to the above the need to respond to real-time changes from the factory, from the supply chain, and from the marketplace, where legacy software solutions lack the needed sensory data (e.g., scalar measurements, images, and videos) to notice changes inside and outside of these connected systems.

1.2.1 Industrial Multimedia Data

Industrial data can be broadly classified into three main categories according to media types: *signal sensing data, tabular and text data*, and *image and video data.* Signal sensing data contains data collected by sensors, actuators, and controllers, including audio data by sound sensors, motion data by infrared optical motion sensors, and trajectory data by sensors in IoT. Tabular and text data contains data stored in both tabular form, free-text form, and semi-structured form. Tabular data consists of structured information with well-defined attributes and properties like sales records, product parameters, and production logs and spreadsheets, etc. Free-text data does not share a predefined structure and consist of floating textual content like user complaints and suggestions. Semi-structured data consists of free-form text that is marked-up and interlaced with elements and attributes, like product descriptions and maintenance record descriptions. More recently, the application of photographic camera equipment in industry and manufacturing has been generating huge amounts of images and video data, which are used to monitor the operations in factories and supply chains, and to check the quality of products. Video data, when extracted by frame, can also be considered as image data. Therefore, methods for handling both kinds of data share many commonalities to a certain extent.

1.2.2 Dynamics of Industrial Data

A key enabler of smart industrial applications is the acquisition of timely and comprehensive data describing the industrial process at hand. Here we distinguish between data-at-rest and data-in-motion which need to be handled differently. Data-at-rest consists of inactive and historical information stored in spreadsheets, databases, and data repositories. This kind of data is primarily utilized to predict and infer long-term data patterns, which are useful in many industrial applications like performance prediction in product planning [50], manufacturing system design [18], and critical event detection [23] like product failure or manufacturing process overload. Data-in-motion describes the active data generated on-the-fly by sensors, actuators, and controllers. This can represent up to 95% of all data generated in a smart manufacturing scenario [51]. This data is constantly generated and ingested into the system to produce a real-time response from the environment. Both data-at-rest and data-in-motion are needed to train and maintain machine learning models and monitor real time condition information such as continuous system diagnosis, prognosis [15], maintenance, communication and collaboration [30] with other related systems.

1.2.3 Processing Multimedia Data

Traditionally, the analysis of industrial data, namely visual data including images and videos, has required human intervention especially during annotation and model training. These tasks are labor-intensive and the balance between efficiency and accuracy is not easy to maintain. Yet with the improvement of computing power and GPU[6] performance, the analysis and processing of images and videos have become more convenient and have established an emerging research direction in manufacturing multimedia data. Here, we distinguish between two types of data processing models in industrial applications: i) batch processing which is adapted for data-at-rest, and ii) stream processing which is adapted for data-in-motion. Using batch processing, the data is first collected in a database or data repository over a period of time, and it is subsequently fed into the processor module for analysis. In other words, we collect a batch of information, and then we process it. Using stream processing, data is acquired and fed into the processor module on-the-fly, piece-by-piece, as it is being collected from the sensors and actuators, where the processing is done gradually in real-time. Coupled with the rise and integrated usage of *big data*, *cloud computing*, and *artificial intelligence* technologies in recent years have made it possible to process continuous streams and large datasets of image and video data more efficiently and effectively.

1.2.4 Need for Synthetic Industrial Data

In addition to processing the real data captured from the physical world, leading manufacturing companies like BMW Group have highlighted the need to generate synthetic image and video data through digital simulation and digital twin technologies, for use in virtual reality (VR) and augmented reality (AR) applications. A digital twin refers to the creation of a digital simulation of a given physical system (e.g.,

---

[6] Graphical Processing Unit.

Fig. 1.5 Extracts of the SORDI dataset: (**a**) First SORDI asset: Smart Transport Robot (iw.hub), and (**b**) Snapshot of the digital twin environment of the BMW factory at Regensburg

creating a digital twin of a physical supply chain or a car manufacturing plan). The digital twin visualization system requires synthetic visual assets that simulate and match their physical counterparts. These assets span from creating a 3D point cloud visualization of a tiny bolt (i.e., the first digital asset in BMW's synthetic industrial assets dataset called SORDI[7]), to digitizing the whole BMW Regensburg plant including its approx. 45,000 square meter work area and more than 35 object classes and approx. 1500 instances labeled in the factory (e.g., pallets, stands, forklifts, and automatic mobile robots or AMRs, cf. Fig. 1.5). More importantly, the synthetic data can be used for augmenting the real data, in order to train more accurate machine learning and computer visions models that operate in both the real (physical) and virtual (digital twin) worlds (e.g., training robots to perceive and understand their visual surroundings, e.g. locating and identifying specific objects in a scene).

1.3   From Digitization to Digitalization and Digital Transformation

Digital transformation is the process of using digital technologies to enhance the performance of new or existing processes and customer experiences in order to meet business variations and market constraints [21]. Digital transformation is being quickly embraced in industry and manufacturing, either in a wholesale approach or following a piece-by-piece implementation where new digital technologies are incrementally integrated to change different aspects of industrial processes. Here, it is important to distinguish between digitization, digitalization, and

---

[7] Synthetic Object Recognition Dataset of Industry (SORDI), www.sordi.ai. SORDI is described in detail in Chap. 6.

their relationship with digital transformation. *Digitization* refers to taking analogue information and converting it into digital data and documents that can be stored, processed, and exchanged by computer systems. The information itself is not changed nor optimized: it is simply encoded in digital format [10]. *Digitalization* uses digital technologies and digital data to transform business and industrial processes and provide openings for new business opportunities and new income, evaluating, re-engineering, and re-imagining the way business and industrial practices are done [20]. Digitalization falls under the larger umbrella of *digital transformation*, which impact in industry is extensive and often includes improvements in safety, quality, throughput, efficiency, revenue, and sustainability, while reducing manufacturing costs to remain competitive in the marketplace [9].

1.3.1   No Turning Back on Digital Transformation

The impact of digital transformation is massive, and what's most important is that there is no turning back: this transformation must happen to keep-up with growing customer needs and fierce competition. It is referred to as the digital die-off reality [32] and can be used to predict a company's persistence or extinction. Companies who do not follow the digital transformation path will probably be left struggling with revenue, dealing with unproductive people, using outdated equipment, legacy software, and outmoded processes [55]. Nonetheless, success in these transformations is not straightforward. Process transformations are usually hard for a large company, and digital transformations are no exception, and might prove to be even harder, with less than a 30% success rate according to McKinsey [8]. In 2019, despite $1.3 trillion invested in transformation initiatives, more than 70% of the said initiatives did not reach their set goals, even when executed by big companies like Ford Motor Company, General Electric, and Procter & Gamble [27]. Various reasons can be speculated behind such failed transformations, ranging from the absence of a clear strategy, to the lack of well-defined goals, lack of the needed human in the loop and material resources, and failing to emphasise the customer's requirements and needs at the center of the digital transformation exercise [27]. However, many companies with successful digital transformation stories focused around developing tailored employee service-based solutions, adaptable digital work platforms, and smart manufacturing solutions, among others, have shown improved stock prices and impressive growth rates, including a 258% growth within 5 years for Microsoft, a 203% growth within 7 years for Hasbro, and 69% growth within 2 years for NIKE [37].

1.3.2   Emerging Digital Technologies

Various emergent digital technologies have allowed the reimagining of many industries. Digital data is becoming increasingly available from a wide range of operational activities using digital devices, i.e., sensors, act accessing and data-processing technologies in everyday objects and partaking in daily operations; we refer to the network connecting such devices as the Internet of Things (IoT), considered as a foundational digital transformation tool. The IoT integrates various data-accessing and data-processing technologies in cyberspace to perceive real-time changes from the real world using digital sensory tools [4]. With 61% of enterprises showing some level of IoT maturity, the usage of data collected from digitally transformed industries and businesses is showing high rates of success [16]. The digital technology infrastructure provided by IoT is promoting the realization of industrial Cyber-Physical Systems (CPS), integrated physical and engineering systems which are monitored, controlled, coordinated, and integrated with computing and communication systems [33]. In turn, the integration of IoT with CPS is promoting the development of so-called digital twins: virtual replicas of physical entities (e.g., a virtual bolt) or entire physical systems (e.g., a virtual factory) that take real-world data about the physical entity as input and produces simulations or predictions through real-time replication, communication, convergence, and self-evolution [19]. With the integration of IoT, CPS, and Digital Twin technologies, massive amounts of data will be generated by these systems and will have to go through the digital data processing pipeline, including data collection, storage, aggregation, analysis, and sharing, in order to eventually provide decision support and data prediction functionality. This in turn requires powerful processing capabilities as well as sophisticated and smart algorithms to deal with the different challenges of industrial big data, which can be handled through dedicated cloud computing and artificial intelligence solutions respectively.

1.4   The Trinity of Big Data, Cloud Computing, and Artificial Intelligence

1.4.1   Industrial Big Data

As fashionable as it may seem, while the label "big data" seems innovative, yet companies have been collecting time series data from factory floors and field assets for decades. Industrial big data originating from Internet-connected automation equipment and plant floor machinery has prominent and perceptible business value for companies looking to augment and transform data into knowledge and smart insights that produce more business value and improved plant performance. However, much of this data remains unexploited, stuck in separate supervisory control and data acquisition (SCADA) systems that are often inaccessible and unavailable to blend with other relevant business data in order to produce meaningful knowledge and insights. And while factory managers and maintenance employees have long analyzed data from specific plant floor assets, mostly with spreadsheets, this was seldom done with wide business transformation in mind. This is fast changing with the huge possibilities provided by emergent digital technologies and advances in edge and cloud computing, artificial intelligence and machine learning analytics. Such technologies allow manufacturers to transform segregated data and systems into dynamic and smart industrial processes and tasks that can be fully or partly automated and optimized in almost real-time. This allows improving and optimizing operations and performance, including reducing maintenance costs, increasing product quality, allowing near-zero downtime, and introducing new revenue streams through new services [46].

1.4.2 Data Processing Using Cloud Computing

With the increasing number of sensors, controllers, and other industrial devices connected to the Internet, legacy centralized data processing servers become easily overwhelmed with the sheer size of the collected data. Hence, there is an increasing need to move added computing power closer to the sensors, i.e., closer to the edge of the network where data is generated. Bringing computational power capabilities to the edge of the network, e.g., on the factory floor, on the conveyor belt, or at a remote solar energy generation site, allows efficient processing of real-time data about the situation and performance of the industrial component without the legacy latency problems that take place when transmitting data to a central processing server for monitoring, analysis, and automation. Cloud computing on the edge of the network, also referred to as edge computing, becomes of central importance when high-bandwidth connectivity is not constantly available, such as in rural areas or in distant factory sites. Handling part of the data processing on the edge nodes of the network alleviates the data processing tasks to be executed on the inner cloud nodes, empowering the latter to undertake the heavier data crunching analytics and prediction tasks. Edge computing also introduces new use cases such as real-time quality management and predictive maintenance of the edge nodes. In these situations, factory floor equipment and industrial components are recurrently monitored and analyzed at the edge, allowing corrective actions like diagnostic checks and initiating maintenance tasks. Companies often devise their own edge and cloud computing solutions that are catered to their needs, where inner cloud nodes usually deliver added storage space, data aggregation from the edge nodes (e.g., combining data from the factory plant controllers, conveyor belts sensors, vision sensors), computation scalability, and data analytics functionality on the aggregate cloud data.

1.4.3 Data Analytics and Prediction Using Artificial Intelligence

Following big data acquisition and procession, the data needs to be crunched and mined for useful analytics and insights, possibly in real-time, to allow smart decision making and initiate corrective measures when needed. Data analytics allows industries to change from reaction to prevention, predicting the behaviors of various industrial systems accurately, and devising actions plans accordingly. In the manufacturing domain for instance, production costs can be predicted by training dedicated artificial intelligence (AI) algorithms – namely data mining, machine learning, and regression analysis models – on big historical production data [6]. Such algorithms are used to generate analytics allowing to evaluate and predict production pipeline performance [6], scheduling of proactive measures when outages occur [36], post-production product performance [50], as well as energy consumption, carbon footprint, and other relevant industrial KPIs [52]. AI-based analytics solutions can extract useful information that industry managers need to make better decisions. In practice, it is often challenging for decision makers to apply their experience and knowledge in new and changing circumstances. Their experience was obtained under the previous circumstance, which may be different from the current one, hence their knowledge may be out of date. With AI-based analytics tools, industry managers and decision makers can analyze historical data, discover new knowledge, and collect useful and actionable insights to make data-driven decisions [9]. Yet major challenges face the usage of AI-based analytics solutions in industry, including the availability of sizable and reliable data, as well as the availability of computation resources and time, especially on the edge of the network – at the level of the sensors and controllers – in order to perform real-time analytics for streaming data.

1.4.4   Challenges with Industrial Big Data

To apply big data technologies in industrial applications successfully, it is essential to assess the usage of big data technologies in industry from the following three perspectives: industrial data, big data technologies, and data applications in industry. First, industrial data empowers modern companies to adopt new data-driven strategies, allowing the transition from legacy industrial practices to modern digitized ones. Yet, it is impossible to consider one big data solution to fit all industrial use cases since different circumstances present different data issues (data types, data formats, and data sources) and require dedicated solutions to handle. Second, we need to better understand the similarities and differences between industrial big data and typical big data on the Web. The 5Vs characters of big data are widely recognized as challenges, such as volume (data size), velocity (ingesting or processing big data in streams or batches, in real time or non-real time), variety (dealing with complex big data formats, schemas, semantic models and information), value (analyzing data to deliver added-value to some events), and veracity (validate data consistency and trustworthiness) [11]. In general, these big data technologies are intended to address some Vs of big data. Hence, their capabilities need to be carefully studied and analyzed to know which Vs are addressed with industrial big data. Thirdly, gaps of data applications in manufacturing need to be identified by reviewing the capabilities of the traditional manufacturing systems and big data analysis. Since much traditional manufacturing software has been widely used in enterprises, the big data produced by such legacy software can be fed back to the modern AI-driven big data ecosystems for analytics and innovative applications such as prediction, optimization, monitoring, simulation, and virtualization [9]. Hence, these application gaps between legacy systems and modern data-driven solutions need to be carefully evaluated to develop adequate digitalization strategies. In summary, knowing the data requirements of manufacturing applications, understanding the capabilities of big data tools, and identifying the gaps between legacy systems and modern technological capabilities, will help define future directions and new ideas for innovative digital transformation processes and smart industry applications, generating significant economic opportunities through industry digitalization.

1.5   The Contribution of BMW Group's Open Source APIs

The BMW Group develops and publishes a range of AI-based algorithms and software tools on an open source platform: github.com/BMW-InnovationLab. These algorithms are part of various applications in production and logistics that focus on automated image recognition and image tagging. They intend to relieve production line and logistic pipeline employees of monotonous tasks such as checking whether a warming triangle is placed in the right location in the trunk of the automobile. This task is currently executed by a camera and a computer vision software that processes the camera's streaming images on-the-fly and detects deviations from the expected norm [3]. BMW Group is sharing parts of their innovative software for image recognition that has been tried and tested in various real and simulated industrial applications, aiming to create a larger community around the technology. This allows BMW Group to receive support and benefit from the community's feedback in order to further develop the software and its use cases, while focusing on implementing more sophisticated AI-based applications in production and logistics. Users of the algorithms and tools are guaranteed anonymity. Errors in the algorithms can be identified quickly, with automated functions from the platform operator providing support if necessary. In terms of quality assurance, the BMW Group checks all incoming user suggestions before they are used in production tools or shared with the community. This allows the initial AI models and their training data to remain intact, pending expert intervention and approval from BMW Group. Users can also decide whether to make their models and data available and accessible to the community, whether to make them accessible to specific stakeholders only like suppliers and industry partners, or whether to keep them private for personal usage and development.

This initiative is part of BMW Group's vision to engage in full-fledged digitalizing, aiming to simulate its entire production process using its Synthetic Object Recognition Dataset of Industry (SORDI, www.sordi.ai), coupled with different AI tools to control robots and industrial machines, and simulate human workers' behavior in the digital twin [28]. Selected algorithms are available online on the open source platform: github.com/BMW-InnovationLab.

References

1. Computer History Museum (CHM). Accessed in June 2023. *The Punched Card's Pedigree.* https://www.computerhistory.org/revolution/punched-cards/2/4
2. National Institution of Standards and Technology (NIST), *Computer Development at the National Bureau of Standards.* NIST Digital Archives, 2001. https://nistdigitalarchives.contentdm.oclc.org/digital/collection/p15421coll5/id/1386
3. BMW Group Press, *BMW Group shares AI algorithms used in production.* BMW Pressclub Global, 2019. https://www.press.bmwgroup.com/global/article/detail/T0303588EN/bmw-group-shares-ai-algorithms-used-in-production?language=en
4. F. Bonomi, et al., *Big Data and Internet of Things: A Roadmap for Smart Environments.* 2014. 546, doi:https://doi.org/10.1007/978-3-319-05029-4
5. M. Canizo, et al., *Real-time Predictive Maintenance for Wind Turbines using Big Data Frameworks.* International Conference on Prognostics and Health Management, 2017. pp. 1–8
6. S.L. Chan et al., Data-driven cost estimation for additive manufacturing in Cybermanufacturing. J Manuf Syst **46**, 115–126 (2017)
7. E.F. Codd, A relational model of data for large shared data banks. Commun. ACM (CACM) **13**(6), 377–387 (1970)
8. Company, M.a., *Unlocking Success in Digital Transformations.* 2018. https://www.mckinsey.com/capabilities/people-and-organizational-performance/our-insights/unlocking-success-in-digital-transformations
9. Y. Cuia et al., Manufacturing big data ecosystem: A systematic literature review. Robot. Comput. Integr. Manuf. **62**, 101861 (2020)
10. B.J. Daigle, The digital transformation of special collections. J. Library Admin. **52**(3–4), 244–254 (2012)
11. Y. Demchenko, et al., *Addressing Big Data Issues in Scientific Data Infrastructure.* Int. Conf. Collab. Technol. Syst. (CTS'13), 2013. pp. 48–55
12. T. Economist, *The World's Most Valuable Resource Is No Longer Oil, but Data.* 2017. https://www.economist.com/leaders/2017/05/06/the-worlds-most-valuable-resource-is-no-longer-oil-but-data
13. R. Ehrenberg, *Square Pixel Inventor Tries to Smooth Things out.* Wired, 2018. https://www.wired.com/2010/06/smoothing-square-pixels/
14. S. Engineering, *Von Neumann Architecture.* Accessed in June 2023. https://semiengineering.com/knowledge_centers/compute-architectures/von-neumann-architecture/
15. R. Gao et al., Cloud-enabled prognosis for manufacturing. CIRP Ann. **64**(2), 749–772 (2015)
16. Gartner, Internet of Things: Unlocking True Digital Business Potential. 2020. https://www.gartner.com/en/information-technology/insights/internet-of-things
17. T. Goldschmidt, et al., *Cloud-based Control: a Multi-tenant, Horizontally Scalable Soft-PLC.* IEEE 8th Int. Conf. Cloud Comput, 2015. https://doi.org/10.1109/CLOUD.2015.124
18. P. Golzer et al., Designing global manufacturing networks using big data. Procedia CIRP **33**, 191–196 (2015)
19. M. Grieves, Origins of digital twin concepts, in *Transdisciplinary Perspectives on Complex Systems*, 2016. https://www.researchgate.net/publication/307509727_Origins_of_the_Digital_Twin_Concept
20. W. Hannah, *What Is Digital Transformation? Definition, Examples and Importance.* Zendesk, 2021. https://www.zendesk.com/blog/digital-transformation/
21. H. Harb, H. Noueihed, *Digital Twin's Promising Future in Digital Transformation* (JOUN Technologies, 2020), 15 p
22. P. Hormann, L. Campbell, Data storage energy efficiency in the zettabyte era. Aut. J. Telecommun. Digit. Econ. (2014). https://doi.org/10.7790/ajtde.v2n3.51
23. S.A. Jacobs, A. Dagnino, *Large-scale Industrial Alarm Reduction and Critical Events Mining using Graph Analytics on Spark.* IEEE 2nd international conference on big data computing service applications. (BigDataService'16), 2016. pp. 66–71

24. M. Jaromir, *The Old Kingdom (c.2160–2055 BC).* in *The Oxford History of Ancient Egypt*, ed. by I. Shaw (Oxford University Press, 2003). pp. 83–107. ISBN 978-0-19-815034-3

25. R. Kimball, *The Data Warehouse Toolkit – The Complete Guide to Dimensional Modeling.* 2nd Wiley Computer Publishing, 2002. 464 p., ISBN: 10:0471200247

26. R. Kirsch, SEAC and the start of image processing at the National Bureau of standards. IEEE Ann. Hist. Comput. **20**, 2 (1998)

27. K. Kitani, *The $900 Billion Reason GE, Ford and P&G Failed at Digital Transformation.* CNBC Evolve, 2019. https://www.cnbc.com/2019/10/30/heres-why-ge-fords-digital-transformation-programsfailed-last-year.html

28. W. Knight, *BMW's Virtual Factory Uses AI to Hone the Assembly Line.* Wired, 2021. https://www.wired.com/story/bmw-virtual-factory-ai-hone-assembly-line/

29. A. Kusiak, Smart manufacturing. Int. J. Prod. Res. **56**, 508–517 (2018)

30. H. Lin, et al., A hyperconnected manufacturing collaboration system using the semantic web and hadoop ecosystem system. Procedia CIRP **52**, 18–23 (2016). https://doi.org/10.1016/j.procir.2016.07.075

31. S. Malaika, M. Nicola, *Data Normalization Reconsidered: An Examination of Record Keeping in Computer Systems* (Developer Works IBM Corporation, 2010), p. 32

32. D. Mazzone, *Digital or Death: Digital Transformation – The Only Choice for Business to Survive, Smash, and Conquer* (Smashbox Consulting Inc., 2014), 166 p

33. L. Monostori et al., Cyber-physical Systems in Manufacturing. CIRP Ann. Manuf. Technol. **65**, 621–641 (2016)

34. P. Moorey, *Ancient Mesopotamian Materials and Industries: The Archaeological Evidence.* 1999. 448 p., Eisenbrauns, ISBN-10:1575060426

35. G. Morahan, *The 20 Most Valuable Companies in Market Capitalisation.* Business Plus, 2022. https://businessplus.ie/news/most-valuable-companies/

36. O. Morariu, et al., *Concept of Predictive Maintenance of Production Systems in Accordance with Industry 4.0.* IEEE 15th International Symposium on Applied Machine Intelligence and Informatics, 2017. pp. 405–410

37. B. Morgan, *7 Examples of How Digital Transformation Impacted Business Performance.* Forbes, 2019. https://www.forbes.com/sites/blakemorgan/2019/07/21/7-examples-of-how-digital-transformation-impacted-business-performance/?sh=2bbe2ec951bb

38. D. Mourtzis, E. Vlachou, A cloud-based cyber-physical system for adaptive Shopfloor scheduling and condition-based maintenance. J. Manuf. Syst. **47**, 179–198 (2018). https://doi.org/10.1016/j.jmsy.2018.05.008

39. National Institution of Standards and Technology (NIST), *Product Definitions for Smart Manufacturing.* NIST – Product data and Systems integration, 2022. https://www.nist.gov/programs-projects/product-definitions-smart-manufacturing

40. National Institution of Standards and Technology (NIST), *First Digital Image.* NIST – Mathematics and statistics, 2022. https://www.nist.gov/mathematics-statistics/first-digital-image

41. D. Price, *Infographic: How Much Data is Produced Every Day?* CloudTweeks, 2020. https://cloudtweaks.com/2015/03/how-much-data-is-produced-every-day/#:~:text=Today%2C%20our%20best%20estimates%20suggest,a%20staggering%2018%20zeros

42. E. Pugh, et al., *IBM's 360 And Early 370 Systems* (The MIT Press, Cambridge, MA, 1991). p. 496. http://www.ibm.com/ibm/history/exhibits/storage/storage_3330.html

43. C. Roemer, *What is the History of Digitization.* Aperture, a Kodak Digitizing Blog, 2018. https://kodakdigitizing.com/blogs/news/what-is-the-history-of-digitization

44. M. Roth, *Law Collections from Mesopotamia and Asia Minor* (Society of Biblical Literature, Atlanta, 1995) ISBN 9780788501043

45. M.G. Siegler, *Eric Schmidt: Every 2 Days We Create as Much Information as We Did Up to 2003..* TechCrunch, 2010. https://techcrunch.com/2010/08/04/schmidt-data/

46. B. Stackpole, D. Greenfield, *Big Data.* Automation World, 2022. https://www.automation-world.com/analytics/article/22485289/big-data

47. M. Syafrudin et al., An open Sourcebased real-time data processing architecture framework for manufacturing sustainability. Sustain **9** (2017). https://doi.org/10.3390/su9112139
48. J. Tekli et al., Minimizing user effort in XML grammar matching. Elsevier Informat. Sci. J. **210**, 1–40 (2012)
49. J. Tekli et al., Approximate XML structure validation based on document-grammar tree similarity. Elsevier Informat. Sci. **295**, 258–302 (2015)
50. J.W. Wang et al., Big data driven cycle time parallel prediction for production planning in wafer manufacturing. Enterp. Inform. Syst. **12**, 1–19 (2018)
51. S. Wang et al., Knowledge reasoning with semantic data for Realtime data processing in smart factory. Sensors **18**, 1–10 (2018)
52. Y. Wu, S. Wang, *Streaming Analytics Processing in Manufacturing Performance Monitoring and Prediction*. IEEE International Conference on Big Data (Big Data' 17), 2017. pp. 3285–3289
53. Y. Busnel, N.R., A. Gal, Y. Busnel, N. Riveei, A. Gal, A. Gal,, *FlinkMan: Anomaly Detection in Manufacturing Equipment with Apache Flinkapache Flink: Grand Challenge*. Proceedings of the 11th ACM International Conference on Distributed and Event-based Systems (DEBS' 17), 2017. pp. 274–279
54. I. Yen, et al., *A Framework for IoT-based Monitoring and Diagnosis of Manufacturing Systems*. IEEE International Workshop on Service-Oriented System Engineering (SOSE' 17), 2017. pp. 1–8
55. F. Zhou et al., A survey of visualization for smart manufacturing. J. Vis. **22**, 419–435 (2019)

Chapter 2
Industrial Evolution Toward the Age of Imagination

The fourth industrial revolution, i.e., Industry 4.0, is stimulating industries to modify and upgrade their business models from a seller's market to a buyer's market. This means that products and processes need to be personalized according to the client's needs, delivered in a short period, while achieving similar or lesser costs as seen in mass production [1]. To achieve the anticipated outcomes, industries are capitalizing on new digitalization technologies that promote manufacturing flexibility and automate decision-making processes [2, 3]. From Artificial Intelligence (AI) and Mixed Reality (MR), to Cyber-Physical Systems, the Internet of Things (IoT), Cloud Computing, and Big Data Analytics, the onset of the Industry 4.0 is gradually altering the workforce skill sets needed at almost all levels of industry. While early technological revolutions focused on the transformation of materials and energy, including water, steam, combustion, and electricity, the current paradigm emphasizes the transformation of information. Less than 1% of the world's information was in digital format in the late 1980s, surpassing more than 99.9% by 2022. Moving forward, Web data statistics show that every 2.5 to 3 years, humanity is producing more information in digital format than since the beginning of human civilization [4]. In this context, the Industry 4.0 also focuses on smart software that can efficiently process digital information and automate its conversion into useful insights and actionable knowledge. This, in turn, is laying the foundation for an upcoming Industry 5.0, where robotic support and human labor work together in continuous cooperation and the human component is considered as the "Centre of the Universe" [5].

***

Then again, how did we reach the fourth industrial revolution? How are we headed to the fifth industrial revolution? What are the prominent digitalization and smart technologies that have stimulated this evolution?

***

We attempt to answer these questions in this chapter and in the remainder of this book…

2.1   From the Industrial Age to the Information and Imagination Ages

Through the course of history, there have been regular advancements in the pursuit of improved and faster production. Among the steady evolution in the industrial progress curve, a few outstanding bends have led to unprecedented enhancements in industrial processes, transformation of various production aspects, and most importantly lasting improvements on humanity's overall quality of life [6]. More importantly, history has taught us thus far that there is not turning back. Failure to accommodate and embrace new technologies in due time has caused major economic complications to industries and has rendered them obsolete (e.g., from manufacturing horse wagons to producing combustion engine cars in the 1920s, and from combustion engines to producing hybrid electric vehicles in the early 2020s). We refer to these outstanding bends in the industrial progress curve as the "industrial revolutions".

2.1.1   The First and Second Industrial Revolutions

The first industrial revolution (1760–1850) essentially had a mechanical theme, revolving around the usage of coal and the creation of mechanical tools made from iron for agriculture, textile looms, and steam-powered locomotives for transportation [7]. This transition to mechanical manufacturing processes took place primarily in England, continental Europe, and the United States, where many of the technological innovations were of British origin [8].This contributed to Great Britain's rise to become the world's leading commercial nation by the late 1700s, creating a global trading empire from North America and the Caribbean to the Middle East and the Indian ocean. The first industrial revolution manifested a major turning point in human history, comparable only to the embracing of agriculture and the shift from nomadic to sedentary lifestyle 4000 years earlier [9]. Average income and population began to show unparalleled and continued growth, where the standard of living for the general population in the Western world began to improve steadily for the first time in history [10].

The second industrial revolution (1880–1973) saw the upgrade of raw materials from iron to steel, and the upgrade of energy sources from coal to refined crude oil. With the advent of electricity, the electrification of cities and industries became a norm. Steam engines were gradually replaced by internal combustion engines, culminating with the booming of the automotive industry in the early twentieth century, marked by a transition of technological leadership from Britain to the United States and Germany. Giant industrial firms started taking form – joining the world's stock markets – like U.S. Steel, General Electric, Standard Oil and Bayer AG. The Toyota Production System (TMS) was introduced to organize manufacturing and logistics for the automobile manufacturer, aiming for seamless production according to a lean manufacturing process to reduce time from manufacturer to clients while minimizing inconsistencies and waste.

This entailed performing special trainings for the human workforce to switch jobs seamlessly within the manufacturing pipeline in order to adapt to the clients' demands and compensate for the gaps [6]. In the meantime, following the creation of the dual-sided through-hole Printed Circuit Board (PCB) design in 1947 [6], the first (Bipolar Junction) Transistor (i.e., BJT) was conceived in 1948 ushering-in the era of integrated electronics and paving the way for the third industrial revolution.

2.1.2   The Third Industrial Revolution and the Age of Information

The third industrial revolution, also known as the digital revolution, highlights the shift from mechanical and analogue electronic technology toward digital electronics which started in the second half of the twentieth century. It is characterized by the production and increased adoption of digital computer systems and digital record-keeping tools, coined with digital computing and communication technologies of that period. Essential to this revolution, is the large scale production and extensive usage of digital logic, Metal Oxide Semi-conductor (MOS) transistors, integrated circuit (IC) chips, and their derivative technologies, including microprocessors, digital cellular phones, and the Internet network infrastructure [11]. These technological innovations have also transformed traditional production and business processes. Computerized assembly lines and terminals were introduced in major factories, coupled with the reduction in size of enormous electrical circuits and their gradual replacement with electronic circuitry. Multiple initiatives were launched to set-up smart industries, transforming the traditional industrial workflow to adopt digitalized designs with smart and interconnected flow lines [12].

Software applications and operating systems also made a gradual entry into the market, leading to a new software development market segment. Computer-aided design software was introduced in the automotive industry in the early 1960s, when IBM delivered its Design Automated by Computer (DAC-1) software for General Motors, to design consumer cars. DAC-1 was run on an IBM 7090 mainframe computer with a dedicated graphics console, where users could draw with a light pen, creating, rotating, and manipulating car images and designs [13] (cf. Fig. 2.1). With the proliferation of personal computers in the 1980s, the software product segment brought about a demand for user data in an effort to provide more personalized services and experiences. This helped launch the so-called Information Age (also known as the Computer Age, or the Digital Age), a period that started in the early 1960s, with a rapid shift from legacy industries, as known during the (first and second) industrial revolutions, to an economy focused on information technology [14]. Industry gradually became more information-intensive while less labor- and capital-intensive, motivating workers to become more productive as the value of their labor decreased [15]. This also marked a shift in the nature of the workforce, emphasizing the need for more technical and design innovation skills to keep-up with the increasing demands and sophistication of the new computerized industry.

Fig. 2.1 DAC-1 running on an IBM 7090 mainframe console [18]

The early 1990s also saw the commercialization of digital networking with the emergence of Internet Service Providers (ISPs). The spread of free information from the Web disrupted the global market dynamics. The Web gradually featured online shopping among other Web-based software applications, highlighting the need for new and adapted logistics. Pizza Hut was the first to jump-in on the Web online shops' bandwagon, and Intershop AG, a German commerce company in 1995 became the online retail goods shopping platform, followed by Amazon and eBay in the same year [6]. The era of Internet-based services had just started. Many logistics firms started delivering on-demand and gradually transformed into more online services. In 1998, Google was launched as one of the first website indexing services. Instead of memorizing the Web addresses of online shops (in the form of Uniform Resource Locators or URLs), Web search engines like Google allow users to easily locate Web addresses using the simple keyword search paradigm [16]. Advertising companies also took advantage of this trend, benefiting from the huge cyberspace coverage provided by the Web as an open announcement and promotional area. In the early 2000s, more explorations into the usefulness of the Internet conveyed one of the key technological concepts of the fourth industrial revolution: the *Internet to control hardware*. This concept was dubbed the Internet of Things (or IoT) by Kevin Ashton, executive director of Auto-ID Center[1] [17].

---

[1] The Auto-ID Labs network (initially founded as Auto-ID Center) is a research group in the field of networked radio-frequency identification (RFID) and emerging sensing technologies. It consists of multiple research universities who initiated the design the architecture for the IoT, and laid much of the groundwork for the standardization of RFID technology and the introduction of the Electronic Product Code (EPC). It continues to research the evolution and application of RFID systems, as well as other disruptive IoT technologies.

2.1.3   The Forth Industrial Revolution Unfolding

The Internet of Things (IoT) is considered as the first pillar of the forth industrial revolution (Industry 4.0). The realization of the IoT vision requires the integration of technologies from various domains to merge and create novel manufacturing processes. Its principal goal is to enable the reception of sensor data from the production floor and generate remote commands accordingly. This technology concept is gradually making it into the market as an e-service for transmitting user sensory data and acquiring user feedback efficiently and securely to control power outlets and appliances of smart homes. In parallel, the automotive industry has been embedding sensor data components of their own to provide innovative features in cars. With more IoT applications and the increased user pool came more sensor data, which – along with social Web data from platforms like Facebook, YouTube and Twitter, would usher in the second key force of the Industry 4.0: Big Data. Meanwhile, the work on automating certain tasks has been advancing in the academic sector with numerous Machine Learning (ML) algorithms designed to solve specific problems. Artificial Intelligence (AI): the third major pillar of the forth industrial revolution, which was first investigated toward the beginning of the cold war, started gaining unprecedented traction in the 2010s with the development of deep learning neural networks to perform sophisticated computer vision tasks that rival human intelligence. This came hand-in-hand with the increased availability of data processing capabilities, providing powerful computing platforms that allowed the efficient training and execution of these deep neural models. In addition to powerful processor architectures, Cloud Computing is considered as the forth pillar of the Industry 4.0, providing enterprise opportunities centered around the provision of high computing resources to clients who could not afford them but requited them for a certain period of time, on-lease following a pay-as-you-go model, for faster and more complex data computations tasks.

The digitalization of industrial production processes, which had started with the third industrial revolution, now has the opportunity to be completely controlled via software. The control systems of PLCs (Programmable Logic Controllers) and the SCADA (Supervisory Control and Data Acquisition) systems, which had been developed since the 1970s, are currently undergoing complete digitalization [6]. In addition, pervasive technologies like Augmented Reality (AR), Virtual Reality (VR), and Digital Twins are attempting to achieve higher levels of integration of cyber–physical and visual enhancements, allowing users to interact with their environment in a pervasive, smarter, and more informed way. These technologies allow users to augment their physical reality (e.g., providing indications and guidance to workers in a factory, highlighting the locations of tools, pinpointing danger zones, etc.) and replicate it in a virtual world (e.g., producing a complete virtual replica of a factory in of the form of a digital twin), helping in simulation, design, navigation, optimization, and prediction, among other information-hungry and visual data crunching tasks. Today, state of the art SCADA systems allow the entire distributed chains of industrial production units to be managed digitally and remotely from around the world, and to be automatically controlled using AI-enabled processes acting on statistically analyzed big data analytics generated from both the real environment and its synthetic virtual replicas. In summary, the forth industrial revolution is currently unfolding, and its technologies highlight a battery of variables and parameters that need addressing and stabilization. While computer-based automation is solving many problems and optimizing many processes, nonetheless, it has also introduces major challenges regarding the coordination and collaboration with non-computerized systems, i.e., the human labor force. The many computerized terminals and robots replacing humans in industry are ushering-in a future where humans and computer systems seamlessly collaborate to achieve common tasks, also referred to as the fifth industrial revolution.

2.1.4   The Fifth Industrial Revolution Toward the Age of Imagination

The goal of the fifth industrial revolution (Industry 5.0) is to establish an industrial system where collaborative robotics (co-bots) and human employees work together in harmony, and where production is highly tailored for the consumers' needs with every component of the product having maximum customizability [19]. This requires technical advancements at both hardware and software levels, assuming that the Industry 4.0 has been successfully realized. The fifths revolution foresees advancement in various aspects, from intelligent manufacturing, to advanced robotics, medical healthcare, and the overall improvement of quality of life. With big data being gradually incorporated into AI-enabled manufacturing, production units would be able to perform analytics automation: pre-ordering input resources from suppliers, keeping inventory, and managing the supply and demand cycle based on consumer demands. Accordingly, a fiscal evolution could follow where automated production unites would be considered as open civil services and would be run as enterprises employing maintenance personnel only [6]. Industrial bots would have all sorts of sensors and actuators and the needed on-board processing power to behave autonomously for long periods of time. Modular bots would have the capability to interface with various standardized components based on industry requirements. This allows producing compound bots, by integrating multiple modular ones, in order to handle special jobs for specific uses cases (e.g., carrying a heavy load might require the combination of multiple modular forklift bots to get the job done, then the individual forklift bots go their separate ways). Human workers would also experience the world with artificially enhanced sensory modules capable of enhancement sensations (e.g., using AI-enabled AR glasses to improve vision on the factory floor, by recognizing objects and tools, pinpointing zones of interest to the individual worker, and providing augmented tutorial trainings to the worker to complete certain tasks). Also, workers would utilize smart tools allowing more precision in executing their tasks (e.g., using smart gloves, connected with the AR-based and AI-enabled systems, to smooth-out the worker's hand movements, reduce quivering, delimit movement within certain boundaries, and guide the worker in achieving critical tasks).

AI and machine learning would handle most of the repetitive and labor-intensive tasks, allowing employees to focus more on creativity and imagination as the primary makers of economic value. In this regard, Industry 5.0 will help usher-in the so-called Age of Imagination, where the rise of immersive reality through combined AI and mixed reality (MR) technologies promise to increase the value of "imagination work" performed by designers, engineers, and skilled workers. Building on the realization of the Age of Information where the focus was on large-scale data processing and data crunching analytics, the Age of Imagination is characterized by intuition and creative thinking to create new user experiences and values. Whether it be creating new industrial processes to optimize production, creating new medical procedures to improve medical surgery, or inventing new processes focused on alternative energy sources, the rise of immersive reality would allow designers and engineers to imagine new solutions, implement them in the virtual world, test them in the virtual world, and simulate their impact and sustainability before ever deploying them in the real world.

Yet while it sounds extremely exciting from a technical perspective, the Industry 5.0 might also introduce some non-technical challenges in human society that need to be addressed.

2.2   With Great Technologies Comes Great Responsibility

2.2.1   Transforming Jobs and Creating New Opportunities

Automation and digitization through computer-based solutions have widely resulted in increased productivity in manufacturing [20], however, Industry 4.0 solutions have also been coupled with human labor loss. In the USA for instance, the number of employees with manufacturing jobs fell from around 17.5 million to 11.5 million from 1972 to 2010, while manufacturing value rose by 270% [21]. Although it initially appeared that job loss in the industrial sector might be partly counterbalanced by the rapid growth of jobs in information technology, yet the recession of March 2001 foretold a sharp drop in the number of jobs in the sector. This pattern of decrease in jobs would continue until 2003 [21], yet data from the past 140 years has shown that, overall, technology creates more jobs than it destroys [22]. A study by the Deloitte consultancy on the relationship between jobs and the rise of technology for England and Wales going back to 1971 shows that technology has gradually taken over hard, dull, and dangerous jobs, which have been counterbalanced by rapid growth in the caring (healthcare and social care), creative, technology, and business sectors. For instance, back in 1971, 6.6% of the labor force in Britain were worked in the agricultural sector – in 2015, this has fallen to 0.2%, a 95% decrease in numbers. During the same time period, there has been a 909% increase in nursing auxiliaries, 580% increase in education assistants, and 183% increase in community workers, among others [22]. In this context, the World Economic Forum estimates that by 2025, technology will produce more than 12 million more jobs than it destroys [23], a sign that automation and digitalization will have a positive impact on the work force and on society.

Yet job creation is not the whole shebang. Equally important is what employees can earn for working those new tech jobs. Do wages rise or fall owing to automation and digitalization? Conventional economics teach us that wages are controlled by supply and demand. When jobs need special skills, wages tend to rise since fewer people meet the demand for these skills. Wages also rise when employees are rare since there are fewer workers available to supply the needed employment. In other words, highly skilled and tech-savvy workers will probably find more job opportunities and demand higher wages, compared with their tech illiterate counterparts. Take idealworks for instance, the wholly owned subsidiary of BMW Group developing AI-enabled logistics solutions. In recent years, the company's autonomous transport bots have become more and more visible in multiple BMW Group plants and at additional customers from different industries. Yet skeptics have raised legitimate concerns about the job losses of transportation employees and jobs in the logistics supply chain in general. Idealworks response: train for and embark on new jobs created by the smart transport logistics industry, like fleet coordination technicians, robot support operators, computer specialists, robotics engineers, and software engineers. "We started back in 2020 with only 9 founding employees from BMW Group, and now we have grown along with our tech supply and support partners to a labor force of more than 100 engineers, technicians, operators, and business officers. Numbers show that we are a job creating company!" says Jimmy Nassif, CTO of IDEALWorks GmbH.

2.2.2   Technology Can Also Reduce Wages

Despite the positive prospects highlighted previously, yet we should add a word of caution: automated systems and robots may sometimes dampen wages and reduce job opportunities. In a recent study published by the American Economic Association [24], the authors found that employees displaced from their jobs due to digitalization and automation are oftentimes required to contend with other employees for whatever job positions are left. For instance, clerical employees who have been replaced by automation may subsequently seek employment in sectors that have not been automated, say retail work. Their entry into the retail sector causes wages in this sector to drop as clerical and retail workers challenge one another for employment [25]. A study by Ashley Nunes of the Harvard Business Review in [25] typically describes this phenomenon. She provides two descriptive cases of the disruptive impact of adopting new technologies. The first case related to aviation. During the initial days of commercial flying in the early tenth century, night pilots commanded higher salaries than day pilots since flying at night was considered to be more dangerous, requiring special skills that were in short supply. Yet as advanced and digitized air traffic control systems become more reliable and cockpit displays became more precise, the risk associated with night flying lessened, along with the necessity of special skills. This eventually led to the elimination of the skills-based wage difference for pilots. Another case described by Nunes is that of taxi drivers in London. They used to ask for a substantial wage premium due to the difficulty of navigating London. They needed to acquire comprehensive mastery of the street maps of the city and few drivers could acquire such knowledge. Yet when Uber started equipping its drivers with powerful smart phone applications that offer turn-by-turn instructions on how to navigate the streets, it eliminated the need for specialized knowledge, and led to reduced taxi driver wages. Nonetheless, it also opened many new job opportunities in the form of Uber drivers.

2.2.3   Need to Integrate New Technologies with Caution

The upcoming Industry 5.0 will focus on collaboration between humans and robots, especially in the industrial sector. Robots will most likely occupy the repetitive and dangerous tasks while knowledge-intensive and creative tasks are handled by humans, together with the responsibility to supervise and monitor the automated robots for maintenance and quality assurance [6]. In other words, new digitalization and automation technologies will not eliminate the need for human labor, it will rather modify the type of work needed. After all, autonomous does not mean human-less [25]. The communications, transportation, commerce, and manufacturing sectors will be first and most affected [26]. The so-called Age of Imagination will leverage new digitalization technologies to allow for enhanced design and innovation. Prototypes and factories will be developed and tested virtually using mixed reality and digital twin technologies. Worker performance will be predicted, and the impact of automation robots will be measured way before building the physical factory or the physical prototype. This is the case of BMW Group's Regensburg factory which was virtualized completely from scratch (cf. Fig. 2.2). This allowed BMW logistics engineers to study the factory's work plans, and optimize its processes, implementing, testing, and improving their thoughts in the virtual environment with minimal cost and human labor. Here, the new digitalization technologies are not replacing the BMW Group factory worker, but are rather augmenting the factory with symbiotic assistive technologies allowing to optimize the workers performance. In this context, issues that might arise from new technologies can be resolved if technological advancements are designed to be human-centric [27]. Developing new digitalization and automation technologies while paying notice to the human-centric factor promises to mitigate the repercussions of adopting those technologies, while focusing on enhancing and transforming human labor versus replacing it. Also, new technologies need to be carefully scrutinized before integration in industry. Executives need to wisely evaluate the potential and the limitations of existing

Fig. 2.2 Snapshot of the digital twin environment rendering of the BMW factory at Regensburg

technologies. What is the technology good for? What can't the technology do? What is the cost and the impact of the technology once integrated in industry? Are the cost and the impact of technology acceptable to stakeholders, including business owners and workers? Executives need to ask themselves these questions when making strategic technology integration decisions [25] to optimize the usage of tech solutions while keeping them under close control and continuous monitoring.

References

1. A. Ciortea, et al., *Industry 4.0: Repurposing Manufacturing Lines on the Fly with Multi-agent Systems for the Web of Things.* 17th International Conference on Autonomous Agents and Multiagent Systems (AAMAS), 2018. pp. 813–822
2. H. Lasi et al., Industry 4.0. Bus. Inf. Syst. Eng. **6**, 239–242 (2014)
3. B. Motyl et al., How will change the future engineers' skills in the Industry 4.0 framework? A questionnaire survey. Procedia Manuf **11**, 1501–1509 (2017)
4. M. Hilbert, Digital technology and social change: the digital transformation of society from a historical perspective. Dialogues Clin. Neurosci. **22**, 2 (2020) https://www.tandfonline.com/doi/full/10.31887/DCNS.2020.22.2/mhilbert
5. V. Martynov, et al., *Information Technology as the Basis for Transformation into a Digital Society and Industry 5.0.* Proceedings of the 2019 IEEE International Conference Quality Management, Transport and Information Security, Information Technologies IT and QM and IS, 2019. https://doi.org/10.1109/ITQMIS.2019.8928305
6. A. Duggal et al., A sequential roadmap to Industry 6.0: exploring future manufacturing trends. IET Commun., 2022. 16:521–531
7. Britannica, *Industrial Revolution: Definition, History, Dates, Summary, and Facts.* Britannica. Accessed June 2023. https://www.britannica.com/event/Industrial-Revolution
8. A. Wrigley, Reconsidering the industrial revolution: England and Wales. J. Interdiscip. Hist. **49**(01), 9–42 (2018)

9. D. North, R. Thomas, The first economic revolution. Econ. Hist. Rev. **30**(2), 229–230 (1977) Wiley on behalf of the Economic History Society

10. R. Lucas, *Lectures on economic growth* (Harvard University Press, Cambridge, 2002), pp. 109–110

11. I. Bojanova, The digital revolution: what's on the horizon? IT Prof. **16**(1), 8–12 (2014)

12. D. Zakoldaev et al., Modernization stages of the Industry 3.0 company and projection route for the Industry 4.0 virtual factory. IOP Conf. Ser. Mater. Sci. Eng. **537**(3), 1–6 (2019)

13. Computer History Museum (CHM), *Computerizing Car Design: The DAC-1.* Accessed in June 2023. https://www.computerhistory.org/revolution/computer-graphics-music-and-art/15/215#:~:text=In%20the%20late%201950s%2C%20the,for%20General%20Motors%20in%201964

14. M. Castells, *The Information Age: Economy, Society and Culture* (Blackwell, Oxford, 1996) ISBN 978-0631215943

15. A. Cooper et al., Initial human and financial capital as predictors of new venture performance. J. Bus. Ventur. **9**(5), 371–395 (1994)

16. J. Tekli et al., Combining offline and on-the-fly disambiguation to perform semantic-aware XML querying. Comput. Sci. Inf. Syst. **20**(1), 423–457 (2023)

17. Postscapes, *Internet of Things (IoT) History*. 2019. https://www.postscapes.com/iot-history/

18. C. Tales, Twitter, 2023. https://twitter.com/computertales/status/863626667910197248

19. German Federal Ministry of Education and Research (BMBF), *Industrie 4.0*. 2016. https://www.bmbf.de/de/zukunftsprojektindustrie-4-0-848.html

20. V. Potocan et al., Society 5.0: balancing of Industry 4.0, economic advancement and social problems. Kybernetes **50**, 794–811 (2020)

21. F. Smith, *Job Losses and Productivity Gains*. Wayback Machine – Competitive Enterprise Institute, 2010. https://cei.org/blog/job-losses-and-productivity-gains/

22. K. Allen, *Technology Has Created More Jobs That It Has Destroyed, Says 140 Years Data Census*. The Gardian, 2015. https://www.theguardian.com/business/2015/aug/17/technology-created-more-jobs-than-destroyed-140-years-data-census

23. M. Kande, M. Sonmez, *Don't Fear AI. It Will Lead to Long-Term Job Growth*. World Economic Forum, 2020. https://www.weforum.org/agenda/2020/10/dont-fear-ai-it-will-lead-to-long-term-job-growth/

24. D. Acemoglu, P. Restrepo, *Automation and New Tasks: How Technology Displaces and Reinstates Labor*. American Economic Association, 2019. https://www.aeaweb.org/articles?id=10.1257/jep.33.2.3

25. A. Nunes, Automation doesn't just create or destroy jobs—it transforms them. Harv. Bus. Rev. (2021) https://hbr.org/2021/11/automation-doesnt-just-create-or-destroy-jobs-it-transforms-them

26. D. Paschek, et al., *Industry 5.0: The Expected Impact of Next Industrial Revolution*. International Conference on Management Knowledge & Learning, 2019. https://ideas.repec.org/h/tkp/mklp19/125-132.html

27. F. Longo et al., Value-oriented and ethical technology engineering in Industry 5.0: a human-centric perspective for the design of the factory of the future. Appl. Sci. **10**(12), 1–25 (2020)

Chapter 3
Background and Technologies

Artificial Intelligence (AI), Mixed Reality (MR), Cyber-Physical Systems, the Internet of Things (IoT), Cloud Computing, Big Data Analytics, and the Digital Twin paradigm are some of the main technologies that have ushered in the fourth industrial revolution and that are stimulating the fifth industrial revolution. They range from smart software that can efficiently process digital information and automate its conversion into useful insights, to predictive and immersive tools that allow robots and humans to work together in continuous cooperation and synchronization, realizing the vision of the industrial metaverse.

***

What are these prominent digitalization and smart technologies about? What problems and use cases do they address? How do they work? And how do they shape the industry of the future?

***

We attempt to answer these questions in this chapter and in the remainder of this book…

3.1 Artificial Intelligence

Artificial intelligence (AI) is a field of study bridging computer science and computer engineering, aiming to simulate biological intelligence and create intelligent entities capable of perceiving their environment and making decisions to maximize their chances of achieving the desired goal, given time constraints and restrictions in processing resources, while resisting noise and uncertainty, making decisions

© The Author(s), under exclusive license to Springer Nature Switzerland AG 2024
J. Nassif et al., *Synthetic Data*, https://doi.org/10.1007/978-3-031-47560-3_3

based on incomplete, inaccurate, or partially incorrect data [121]. Creating synthetic intelligence comes down to combining and integrating a number of technologies and techniques from different fields, including search and planning, knowledge representation, natural language processing, machine learning, social intelligence, and generative computing.

3.1.1 Search, Planning, and Motion

Search agents allow solving a problem intelligently by searching for the best solution in a space of many possible solutions. This requires producing an abstract mathematical representation of the environment called the state space (i.e., the set of states an agent can find itself in), identifying the set of actions an agent can undertake to change states called the transition model, identifying the agent's goal (i.e., the state it needs to reach representing the solution to the problem being formulated), and identifying the agent's utility function to evaluate the quality of a solution. The agent is supposed to reach an optimal solution, i.e., the most cost effective solution under given constraints. Typical applications of search agents include intelligent navigation systems (e.g., car navigation using GPS, robot navigation within a factory plan), planning and scheduling (e.g. from logistics and mission planning [38, 66] to health monitoring and meal plan recommendation [125, 126]), gaming and simulation (e.g., bot chess players, maze and puzzle solvers [110, 121]). Search-based solutions like SLAM (Simultaneous Localization And Mapping) are at the core of motion planning, environment planning, localization, and navigation solutions for robotics and assistive systems and applications [136], such as automating the movement of ground robots inside a factory, coordinating the movement of drones in remote natural areas to perform air pollution monitoring, gas leakage detection, and power grid failure, prompting quick actions accordingly to minimize damage to the environment.

3.1.2 Knowledge Representation

Many of the tasks that machines are expected to handle require knowledge about their environment and the world they operate in. AI needs to represent objects, categories, actions, events, their properties and relationships. This allows associating the available information with well-defined meaning, to be analysed and processed by machines [111]. This includes machine-readable environment, climate, and weather ontologies, such as ENVO: an expressive semantic graph which helps humans and machines understand environmental entities of all kinds, from microscopic to intergalactic scales [156], Also in the Natural Language Processing (NLP) and Information Retrieval (IR) fields, linguistic knowledge bases (such as WordNet [98], Roget's thesaurus [155], and Yago [67]) provide a framework for organizing words/expressions into a semantic space [140]. A linguistic knowledge base is usually represented as a semantic network made of a set of entities representing semantic concepts or groups of words/expressions, and a set of links between the entities, representing semantic relationships (*synonymy*, *hyponymy*, etc.). They provide ready-made sources of information about word senses to be exploited in various language processing tasks including content analysis [133], sense disambiguation [139], document classification [137], image semantization [124], multimedia metadata clustering [123], and collective knowledge management.

3.1.3 Natural Language Processing

Natural Language Processing (NLP) is an interdisciplinary field intersecting linguistics, computer science, and artificial intelligence, concerned with understanding and generating human language using techniques from knowledge representation, semantic/statistical analyses, and machine learning, in order to allow human-machine communication and interaction. Applications of NLP include information retrieval, text miming, lexical and semantic analysis, contextualization, semantic inference, and machine translation, e.g., [57]. The ultimate goal of NLP is to allow a computer to understand the meaning of human text and speech, and allow it to interact with the human correspondent accordingly. In the past few years, spoken dialogue system interfaces have gained increasing attention, with examples including Apple's Siri, Google Assistant, Microsoft's Cortana, Amazon's Alexa, and numerous other products. Most existing solutions utilize deep learning models, where recurrent neural networks (RNNs) have been successfully adapted to dialogue systems through encoder-decoder architectures [95, 148]. More recently, OpenAI's ChatGPT took the world by storm, at the next generation generative transformer-based model. Compared with Siri and Google Assistant which are virtual assistants that respond to natural language commands to perform tasks (e.g., control Apple devices, or search for information on the Web), ChatGPT is a full-fledged chat bot designed to perform human-like conversations and solve problems through textual conversations. Having reached human-like accuracy levels, dialogue systems are being increasingly investigated in different fields including industry and manufacturing, where NLP helps automatically analyze notes and texts in manufacturing and client records [70]: extracting useful insights from these texts, and performing human-like conversations with employees and clients allows improving manufacturing processes and customer service.

3.1.4 Machine Learning

Machine Learning (ML) Refers to the science of getting computers to perform tasks without being plainly programmed for those tasks, where they evolve their own behaviours based on empirical and historical data from sensors or data repositories [100]. ML aims to automatically recognize complex patterns and make intelligent decisions by learning from and comparing with previous experiences. It allows solving pattern detection and recognition problems, including face recognition, speech recognition, and object recognition [121]. Industrial applications of ML include performing obstacle detection for transport robots in a factory plant, recognizing objects and workers' faces in a plant, detecting energy emission reductions, CO_2 emissions, and air quality monitoring. ML can also help to lessen the frequency of incidents occurring in a factory, by detecting and recognizing patterns and changes in the data, identifying faults in real-time, and adjusting automated decisions accordingly to ensure that such incidents have minimum effects on the factory and the damage they cause [108]. We describe and categorize machine-learning algorithms in more detail in the following section.

3.1.5 Social Intelligence

Refers to the usage of machine-readable knowledge, ML, and language processing to detect and recognize sentiment and emotion patterns in human speech and behaviour, and react accordingly (emotional robots, robotic assistants, chat bots, etc.) [2, 42]. At the core of social intelligence are sentiment analysis tools which analyze words, text, and speech extracts provided by users, and attempt to classify them under different sentiment categories, such as: *positive*, *negative*, or *neutral* emotions. Affect analysis can be viewed as a more fine-grained approach of sentiment analysis, which involves more specific classes of affective emotions such as: *happiness*, *sadness*, *surprise*, and *anger*, etc. Many sentiment analysis approaches utilize ML techniques applied on corpus-based statistics in order to match textual or speech patterns with sentiments represented as labeled categories, e.g., [1, 53]. Other techniques utilize knowledge-based solutions, e.g., [52, 147], in order to match target words with seed words in a sentiment lexicon (e.g., List of Emotional Words – LEW [49], or WordNet Affect – WNA [146]), by evaluating their semantic similarity or distance in a reference lexical knowledge base (e.g., WordNet [98]). Most methods usually produce sentiment labels (e.g., *joy*, *surprise*), while evaluating sentiment intensity (valence) scores (e.g., 20% *joy*, 35% *surprise*). Such solutions are central in developing robot assistants and co-bots in factories which display human like behavior, and are capable of recognizing the emotional states of human employees in order to adapt their communication and interaction accordingly. Such methods can also be used to perform client feedback analysis [46, 145] (automated analysis of customer opinions on purchased products), blog and social media sentiment analysis [43, 157] (analyzing bloggers' reviews in web forms regarding certain products, topics, events, or people).

3.1.6 Evolutionary and Generative Computing

In addition to searching, decision making, noise resistance, and learning, a truly intelligent system should be able to create and pursue its own goals, to persist without (human) assistance, for a long time, and to expand its knowledge beyond predefined constraints. This is where evolutionarily and generative computing comes to play. Inspired by the Darwinian Theory of Evolution, genetic algorithms are the most famous form of evolutionary computing, mimicking genetic processes that nature uses to produce new solutions from existing ones, where the new solutions supposedly surpass their existing predecessors' quality (referred to as solution fitness). Genetic algorithms are probabilistic optimization algorithms using concepts of Natural Selection and Genetic Inheritance. A genetic algorithm maintains a population of candidate solutions for the problem and makes it evolve by iteratively applying a set of stochastic operators. These operators include *selection*: replicating the most successful solutions found in a population at a rate proportional to their relative quality (fitness), *recombination*: (crossover) decomposing two distinct solutions and then randomly mixes their parts to form novel solutions, and *mutation*: randomly perturbing a candidate solution by including controlled errors. Recently, Generative Adversarial Networks (GANs) have been receiving increasing attention as a new family of Machine Learning (ML) algorithms designed to generate new data or solutions based on existing ones. Given an initial population of data items, referred to as training set, GANs learn to generate new data with the same statistics as the training set. For instance, a GAN trained on photographs of items in a factory can generate new photographs of the same items that look seemingly authentic to human observers, having many realistic characteristics, albeit with new properties encoded in the GANs' generative model. This is especially useful when creating a virtual environment or a digital twin representing a factory plant for instance, where items in the virtual world are created synthetically to mimic their real counterparts, albeit with special simulated properties that are induced by the virtual environment designer to study different aspects of the factory production process.

While originally designed to perform generative computing, GANs have also proven useful in different ML scenarios, including supervised, semi-supervised, and reinforcement learning (cf. Sect. 3.2) (Fig. 3.1).

3.2 Machine Learning

3.2.1 Supervised Learning

Supervised learning is subcategory of Machine Learning (ML) that uses labeled datasets to train algorithms to categorize or predict the labels of new data. Labelled data – also referred to as training data, consist of data points/objects with associated

Fig. 3.1 Comparing supervised learning and unsupervised learning (**a**) Supervised learning. (**b**) Unsupervised learning. (Adapted from [91])

labels – referred to as target labels, target categories, or target classes. More formally, a dataset of labelled data can be represented as a set of pairs $\{(X_1, y_1), \ldots (X_n, y_n)\}$ where each pair (X_i, y_i) represents a data point X_i as a feature vector in a feature space, and a label y_i associated with X_i [103]. For instance, X_i could represent the color histogram feature vector describing the image of a forklift, and y_i could be the label *"forklift"* associated with it. The labels provide linguistic and qualitative descriptions of the otherwise purely mathematical and scalar feature vectors [8]. Supervised learning can also be performed in the form of numerical regression: given a set of scalar data point samples, trying to produce a function that generates the outputs from the inputs [6]. Given a large set of labelled data (e.g., a large set of labelled images of *"forklifts"*), the purpose of supervised learning is to learn a function that associates the target labels with new incoming unlabeled data (e.g., given a new image of an industrial truck, the supervised learning algorithm needs to decide whether it will be categorized as *"forklift"* or not). The latter is referred to as *binary classification*: deciding whether one single label is achieved or not. This can be extended to perform *multi-class classification* where the labeled dataset consists of data points associated with different labels, where each data point has one single label only and thus belongs to one single class (e.g., an industrial truck can either be a *"forklift"* or a *"transport truck"*, not both at the same time). Certain application scenarios might also require *multi-label classification* which assigns more than one label to the same data point (e.g., an industrial truck can be labelled as *"forklift"*, *"electric vehicle"*, and *"smart bot"* simultaneously, describing an smart electric forklift robot). Multi-class classification can be considered as a special case of multi-label classification, where the highest membership class is only kept, and the other lower-membership classes are disregarded. Supervised learning algorithms analyze the labelled training data and produce an inferred function, which can be used for mapping new data [121]. An optimal scenario allows the algorithm to correctly determine the target label(s) for new unseen data points. This requires the learning algorithm to train by generalizing from the training data to handle unseen data points which is not always an easy task [6]. Here, we distinguish between parametric learners which require setting some parameters during training and execution, like the number of layers and the number of cells per layer in an Artificial Neural Network (ANN). Tuning a set of fixed size parameters simplifies the learning process, since parametric assumptions remain independent of the number and nature of the training data. Yet a set of fixed size parameters would also limit what can be learned by the ML algorithm, which tends to limit its expressiveness [121]. Parametric learners include ANNs, Bayesian Networks, and Logistic Regression models, among others. On the other hand, non-parametric learners are mostly data dependent and require setting fewer parameters. They are mostly free to learn any functional form from the training data, regardless of specific parametric choices. Nonetheless, they are usually slower to train since they need to consider all the labelled data without parameter simplifications, and they are usually prone to overfitting [6]. Non-parametric learners include K-nearest neighbors, support vector machine, decision tree, and random forest, among others.

3.2.2 Unsupervised Learning

Different from supervised learning, unsupervised learning uses unlabeled data. It aims at finding hidden patterns or structures within the unlabeled data. This is usually performed with the absence of target data labels or knowledge about the labels' existence. In other words, unsupervised learning is conducted without the need for labelled training data, learning without external intervention [62]. Data clustering is the most common form of unsupervised learning, where unlabeled input data points/ objects are automatically organized into so-called clusters of similar or related objects, such that different clusters describe relatively different or unrelated objects. Data clustering algorithms can be organized in three main categories: (i) partitional, (ii) hierarchical, and (iii) other methods. Partitional clustering algorithms attempt to divide data objects into non-overlapping subsets, i.e., the clusters, such that each data object is in exactly one cluster, by maximizing intra-cluster similarity and minimizing inter-cluster similarity. K-means [68] is one of the most popular algorithms in this category and attempts to recursively minimize the distance between objects in a cluster and a special object designated as the center of the cluster (computed as the average between all objects in the cluster). The clusters are re-computed and adjusted recursively until reaching convergence (where the cluster centers remain unchanged). Hierarchical clustering algorithms generate a set of nested clusters organized in a hierarchy, called *dendrogram*, where the root node of the dendrogram represents the whole dataset and each leaf node represents an individual data object.

The cluster hierarchy is produced based on the similarity between individual data objects/clusters. Agglomerative hierarchical clustering for instance starts with each object forming its own cluster, and then finds the best pair to merge into a new cluster, recursively repeating this process until all clusters are fused together. Hierarchical clustering algorithms usually produce better results compared with their partitional counterparts, yet they are usually more computationally complex (they usually required quadratic time, compared with many partitional algorithms like k-means which are of average linear complexity). Also, partitional and hierarchical algorithms tend to break down large clusters into smaller (more homogeneous) ones which might not be always favorable from the user/application's side [141] . Other clustering approaches which are designed to produce contiguous, time-sensitive, or irregular clusters include *incremental*, *density-based*, *spectral*, and *fuzzy* clustering algorithms [60, 158].

3.2.3 Semi-supervised Learning

Semi-supervised learning is a type of ML that deals with a small amount of labeled data and a large amount of unlabeled data during training. The main premise of semi-supervised learning is that some labelled training data is available, yet it is not enough to train the ML algorithm. Hence, there is a need to extend the small training dataset by labelling part of the unlabeled data and integrating it into the labelled training dataset, providing additional training pairs sufficient to train the ML algorithm. One famous semi-supervised learning approach is *bootstrapping* [14, 99], which uses the labelled data to train one or more lean ML models that are simple enough to train on the available training data. For instance, consider a small set of labelled images showing the faces of registered clients for a certain retail shop, where the images are labelled according to gender: *"male"* and *"female"*, in order to promote the shop's merchandise accordingly. Given that the labelled dataset of facial images is not enough to train the initial gender classification model, simple classifiers can be defined to process simple facial features separately, e.g., a simple classifier to distinguish gender according to the position of the eyes, another classifier to process the shape of the eyes, and other classifiers to process the size of the eyes, the shape of the hair, the position of the eye brows, etc. Training simpler classifiers to process simpler features might prove effective with the small training dataset. Once trained, the unlabeled input data (e.g., unlabeled client facial images) are run through the simple models. For each unlabeled data object, if the simple classifier concur on the same output label, then the label is associated with the input data to form a labelled pair, and the pair is added to the training dataset. Otherwise, the input data object is dismissed. This process is repeated on the unlabeled dataset until enough data objects have been labelled by the simple classifiers and the size of the augmented training dataset is large enough to train the initial ML algorithm. Other semi-supervised learning approaches include *graph regularization* methods [81, 159] which build models based on similarity graphs, where nodes in the graphs represent data objects and edges in the graph represent data similarity. A regularization constraint is applied on the graph so that unlabeled nodes which are similar enough according to the graph share similar labels. Another approach is *structural learning* [96, 153] which uses unlabeled data to generate a new reduced-complexity hypothesis space by exploiting regularities in the feature space via unlabeled data. In other words, the initial feature space is transformed into a new one where more unlabeled data can be easily labelled through existing semi-supervised approaches such as bootstrapping or graph regularization described above. This is conducted with the expectation that the transformed feature space allows to produce a good and large enough augmented training set for the initial ML model.

3.2.4 Reinforcement Learning

Reinforcement learning is subcategory of ML where algorithms are designed to reward desired behaviours and punish undesired behaviours. They assign positive values to the desired actions to encourage the ML algorithm, and negative values to undesired behaviours. This allows the model to seek long-term and maximum overall reward to achieve an optimal solution [27] . The long-term goals help prevent the ML algorithm from delaying on lesser goals. With time, the model learns to avoid the negative values and seek the positive ones. This mirrors processes that appear in animal psychology. For example, biological brains seem programmed to interpret signals such as pain and hunger as negative reinforcements, and interpret pleasure and food intake as positive reinforcements. Smart animals like apes, dogs, and humans learn to engage in behaviors that optimize these rewards. A typical reinforcement learning model is modelled as a Markov Decision Process (MDP) which interacts with its environment in discrete time steps [72, 128]. At each periodic time stamp, the model receives the current state and the reward associated with it. It then chooses an action from the set of available actions, which is afterward sent to the environment. The environment moves to a new state and the reward associated where the state transition is computed. The goal of the model is to learn a function which maximizes the overall aggregate reward as the model transits from one state to another, until reaching its final state(s) or until its operation is terminated.

3.2.5 Deep Learning

Deep Learning (DL) is a category of Machine Learning (ML) models based on multi-layered Artificial Neural Networks (ANNs) which are inspired by neural structures in biological systems, albeit with many differences in terms of elementary cell representations, network architectures, and dynamics. A biological neuron (cf. Fig. 3.2a) is a cell consisting of a cellular body and a nucleus. The cellular body contains ramifications: dendrites through which electro-chemical stimuli are transmitted to the neuron, and an axon through which the neuron's electro-chemical response signal is transmitted. At a given time, the neuron receives input stimuli from its neighboring neurons, processes the received signals within the cellular

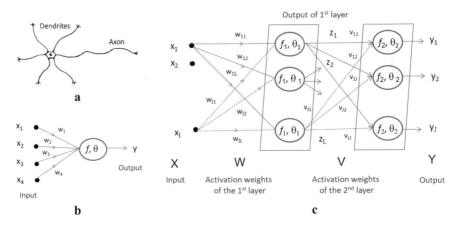

Fig. 3.2c Sample ANN representation. (**a**) Individual neuron, (**b**) Indivitual cell model, (**c**) Two-layered ANN model

body/nucleus, and produces or not an output response given the intensity and frequency of the input stimuli and the sensitivity of the neuron's cellular body. The latter is formally referred to as the neuron's activation function. ANNs attempt to simulate the latter behavior mathematically (cf. Fig. 3.2b), connecting multiple neural cells together in specific ways to produce a neural network (cf. Fig. 3.2c).

DL models use multilayered neural network architectures. The term "deep" usually refers to the large number of internal layers within the neural network. Traditional ANNs typically contain 2–3 layers while deep networks can encompass as many as 150 layers. DL models are trained by using large sets of labeled data and utilize neural network architectures that learn features directly from the data without the need for manual feature extraction [94]. A famous deep neural network is known as Convolutional Neural Network (CNN) which uses 2D convolutional layers, making it well suited to processing 2D data, such as the images' pixel matrices (cf. Fig. 3.3). CNNs extract features directly from images through a sequence of convolution layers performing "end-to-end" learning, compared with typical ML algorithms where the process starts with relevant features being manually extracted from images (cf. Fig. 3.4). Other famous DL algorithms include Recurrent Neural Network (RNN) where the output from certain cells of the output layer feeds back to the hidden layer(s), allowing dynamic temporal behavior for a time sequence. The cells' internal states serve as short-term memory to process sequences of inputs, useful in tasks such as unsegmented handwriting recognition and speech recognition. Transformer models and GANs (Generative Adversarial Networks) are also receiving attention in various application domains, especially in linguistics to perform machine translation and speech synthesis in conversational systems [95]. However, we note that DL algorithms are more computationally complex and typically require huge amounts of labelled data (at least a few thousand data instances) to perform automatic feature extraction and achieve acceptable learning rates, compared with ML algorithms which are relatively less computationally complex and

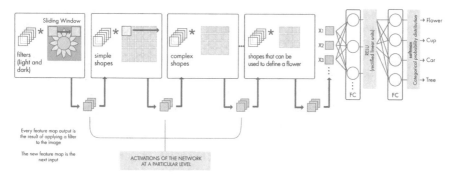

Fig. 3.3 Sample CNN with many convolutional layers, reported from [94]

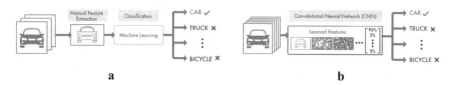

Fig. 3.4 Comparing the ML approach (right) with the DL approach (left), reported from [94]. (**a**) Typical machine learning process. (**b**) Deep learning process

require relatively less training data (unsupervised learners like decision trees can achieve acceptable results with hundreds of labeled data instances [112]). Running DL algorithms often required specialized hardware like Graphical Processing Units (GPUs) and powerful Solid State Drive (SSD) memory cards.

3.3   Computer Vision

Computer Vision (CV) is a subfield of machine learning that trains computers to interpret and gain high-level understanding from visual data [104]. It seeks to automate tasks that the human visual system can do, ranging over the automatic extraction, analysis, and understanding of useful information from single images or temporal sequences of images, i.e., videos [131]. It involves the development of a theoretical and algorithmic basis to achieve automatic visual understanding. The visual data can take many forms, such as individual scenes, views from multiple cameras, or multi-dimensional data from a medical scanner. "Visual understanding" in this context means transforming the visual input into numerical or symbolic information, providing descriptions of the world that lead to appropriate actions or decisions by computer systems. CV solutions are application-dependent: they can be used as standalone solutions that solve a specific problem (e.g., scene reconstruction, 3D scene modelling, object detection, event detection, video tracking, 3D pose estimation, and image restoration [4, 5]), or as part of a larger system including other subsystems for control of mechanical actuators, sensors, and information databases [104].

A typical CV pipeline consists of six main steps. First, *image acquisition* allows acquiring a digital image representation from one or several image sensors, including different types of light-sensitive cameras, range sensors, ultrasonic sensors, and tomography devices in medical applications. The resulting digital image representation depends on the type of sensor, and can range from a typical pixelated 2D image, a 3D volume, or a temporal image sequence (i.e., a video). The pixel values usually represent light intensity in one or several spectral bands forming gray-scale or color images, and can also represent other measures such as depth, absorption of sonic waves, or reflectance of electromagnetic waves [34]. Second, *image pre-processing* is usually applied to improve the digital image representation, including re-sampling, noise reduction, contrast enhancement, light enhancement, and multi-scaling to enhance image structures at locally appropriate scales [34]. Third, *feature extraction* allows extracting appropriate image features, including colors, lines, edges, localized interest points, as well as more sophisticated features related to texture, shape, and motion (we will cover image feature representation in more detail in Chap. 4 of the book). Forth, *image segmentation* is sometimes used alongside feature extraction to identify which regions of the image are more relevant for further processing. This can result in identifying regions containing specific objects of interest, decomposing an image into a nested scene structure including foreground, background, object groups, and salient object parts [13]. Segmentation can also be applied to videos, decomposing a video into a series of per-frame foreground and background masks, while maintaining the video's temporal continuity [87]. Fifth, *high-level processing* is conducted on the extracted image features, to perform a high-level application-specific task ranging over scene reconstruction and estimation of specific parameters such as object pose or object size, classifying a detected object or event into different categories, and comparing and combining different views of the same object. Sixth, the decision making phase allows making the decision needed for the application, such as *pass* or *fail* on automatic inspection (e.g., detecting visual defects in manufactured products, classifying the type of damage), *match* or *no match* in recognition applications (e.g., recognizing specific defects in products, recognizing employees' faces), and flag for further human review (especially in medical and security recognition applications, such as flagging a suspicious tumor-like shape in an organ tissue [80], or flagging suspicious items at an airport [160]).

3.4   Graphics Rendering

Graphics rendering is the process of generating a realistic or non-realistic image from a 2D or 3D model using a dedicated computer software. Models are first defined as wireframe sketches in a virtual scene containing geometry, viewpoint,

a   b

Fig. 3.5 Sample renders based on synthetic assets from SORDI (https://www.sordi.ai) (left), and visualizing BMW Group's Regensburg production line (right) using NVIDIA's Omniverse engine. (**a**) Render of idealworks' iw.hub. (**b**) Render of BMW Group's Regensburg production line

texture, lighting, and shading information describing the scene. The data contained in the virtual scene is then passed to a rendering software which adds a battery of rendering features including textures, lights, shadows, reflections, transparency, refraction, and motion blur, among others [55, 114], to be processed and output as a digital image representation. Powerful rendering software include Unity,[1] Blender,[2] NVIDIA Omniverse,[3] and Adobe Substance 3D,[4] among others. The resulting image is referred to as the *render* and represents a complete image that the intended viewer can see (cf. Fig. 3.5).

Different rendering techniques have been developed in the past few decades. The ideal rendering technique would process every pixel in the scene, which is unfeasible and would require a huge amount of processing time. *Rasterization* for instance produces a high-level representation of the image where specific objects are identified and processed as so-called *primitives*. Instead of performing a pixel-by-pixel rendering, rasterization using scanline rendering follows a primitive-by-primitive approach, looping through the primitives, determining which pixels they affect, and modifying those pixels accordingly [114]. Another rendering technique is *ray casting* which evaluates the scene as perceived from a certain point of view, and calculates the perceived image from the point of view outward, line by line as if casting rays out from the point of view. It calculates the ray of light from the object to the point of view, considering that the light ray follows a straight path [55]. Another calculation is made of the angle of incidence of light rays from the light source(s), considering the source(s)' intensities. A third rendering technique is *ray tracing* which is similar to ray casting, but uses more advanced optical simulations like Monte Carlo methods [93] to obtain more realistic results at a significantly faster pace. More recently, neural rendering has been used to construct 3D models from 2D images using ANNs [142], collecting images from multiple angles of an object

---

[1] https://unity.com/.

[2] https://www.blender.org/.

[3] https://www.nvidia.com/en-us/omniverse/.

[4] https://www.substance3d.com/.

and mashing them into a 3D model. This can be computationally intensive and might require pre-rendering the images offline and then using them when needed. Nonetheless, real-time rendering is sometimes required for 3D game videos and mixed reality simulation environments that need to dynamically create scenes and adapt to the user behavior on-the-fly. In such applications, 3D hardware accelerators can improve real-time rendering by using the graphics processor on the video card (GPU) instead of consuming valued CPU resources [142].

3.5   3D Scanning

3D scanning is the process of examining a real-world object or environment to collect data on its shape and appearance. The data is collected in the form of a polygon mesh or a point cloud, and is processed to extrapolate the shape of the object and construct a digital 3D model of it. The 3D model is then processed for graphics rendering as described in the previous section. Different from regular cameras, 3D scanners collect distance information about the surfaces within its field of view, describing the distance to the surface at each pixel in the produced image. Various types of 3D scanners exist, including optical, acoustic, radar, and laser. In industrial settings, 3D laser scanning has been recently used to capture production facilities' spatial properties in order to produce so-called Virtual Factory Layouts (VFLs) [86]. The 3D laser scanner sends out laser beams that reflect back when they hit objects, at which point the distance the laser beam has travelled is measured [76]. Measurements are stored as points with XYZ-coordinates relative to the scanner-position, creating a cloud of points that visualizes the scanned environment. The point clouds' density depends on the laser scanners' performance and the set resolution [33]. The 3D laser scanner can generally rotate 360 degrees around its vertical axis and 300–320 degrees around its horizontal axis, giving a large field of view. In addition, many 3D laser scanners have digital single-lens reflex cameras that can capture images and map them to the point cloud [39]. This provides each point a particular color producing a more genuine representation of the real environment. Several scans of large areas and several sides of an object can be captured separately, and then merged together to form a large point cloud [16]. Computer-generated (synthetic) objects can be imported into the point cloud, to help realize more sophisticated virtual environments [86].

To make the point cloud more easily editable, the list of points in 3D space can be transformed into a solid model using 3D modelling techniques such as *voxelization* and *high-level 3D object representations*. Voxelization is easily understood when compared with pixels of 2D images. The word *pixel* originates from words *picture* and *element* – similarly, the word *voxel* originates from *volume* and *element*. Raster images are represented as 2D grids where each cell in the grid is called a pixel and can be addressed by its (x, y) coordinates – similarly, voxels represent cells in a 3D grid where the value of a voxel is accessible through its (x, y, z) coordinates. The structured nature of a voxel representation makes it easy to adapt 2D

Fig. 3.6 Generating SORDI 3D point cloud assets using NVIDIA Kaolin

Fig. 3.7 Virtual factory layout of BMW Group's Regensburg plant, using point cloud-based high-level 3D object representations

object recognition methods to 3D space; most notably 3D convolutions. Convolutional Neural Networks (CNNs) are probably the single most powerful methods in 2D computer vision and usually outperform other methods in the field. The success of CNN on 2D images encouraged researchers to apply the same technique on voxel representations. However, voxel representations add processing overhead to convert 3D point clouds into voxels before using voxel-based computer vision methods, and might require hand-crafted features to encode the content of a voxel (e.g., dealing with multiple points located within the same voxel, choosing voxel size/resolution).

More recently, high-level representations like scene graphs have been increasingly used to encode and group geometric shapes into hierarchical structures, showing the benefits of 3D scene reconstruction using object detection in point clouds. For instance, BMW Group and idealworks have adopted Pixar's Universal Scene Descriptor (USD) standard in generating their SORDI dataset 3D assets (cf. Fig. 3.6). USD is also adopted by NVIDIA in their latest virtualization engine Omniverse, which is used by BMW Group and idealworks to develop full-fledged Virtual Factory Layouts (VFLs) using SORDI (cf. Fig. 3.7). High-level representations introduces the concept of 3D objects and are not just a collection of points or voxels. This conceptual idea simplifies many otherwise complicated operations, including object and scene detection, recognition, translation, cropping, scaling,

Fig. 3.8 3D point cloud manipulation: selection, and cropping, and filtering of point cloud layers (based on BMW Group's Regensburg plant VFL model)

and rotation (cf. Fig. 3.8). In addition, high-level 3D representations can be more efficiently rendered and used for physics and motion simulation, which are crucial operations for manipulating VFLs.

3.6   Immersive Technologies

Visualization for replacement in smart manufacturing [161] consists in presenting complex and precarious work scenarios in a Virtual Factory Layout (VFL) where people can train, learn new skills, and practice in a safe and informative environment. This requires 3D scanning and graphics rendering to create the VFL and its visual structures, coined with the usage of *immersive technologies* to bring the virtual environment to life and simulate the behavior of its internal and external components. Immersive technologies refer to software and hardware solutions that blur the boundaries between the physical world and the simulated one [58]. These can be classified under Virtual Reality (VR) and Augmenter Reality (AR).

3.6.1   *Virtual Reality*

Virtual reality (VR) is a set of technologies that enable users to immersively experience a fantasy world beyond the realm of reality [18]. Various VR technologies have been developed over the past years, including different display, interactions, and tracking solutions. Display technologies come in a variety of modalities and sizes, aiming to deliver different kinds of information to the human senses, including sight, hearing, and touch. Visual displays come in different shapes and configurations, e.g., a single large projection screen (i.e., Powerwall [120]), multiple connected projection screens (i.e., CAVE [31]), stereo-capable monitors with desktop tracking (e.g., INL VR [40]), and head-mounted displays (e.g., Oculus Rift [24] and HTC Hive [151], cf. Fig. 3.9). Audio displays stand for headphones, a single speaker, or a full surround sound system. Sound localization makes it possible to simulate

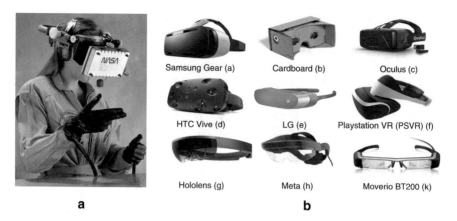

Fig. 3.9 Sample VR head mounted systems (**a**) NASA's 1990 legacy VIEW VR system [107]. (**b**) Sample VR head mounted displays

sound moving or coming from a location within a virtual environment [18]. Interaction with a virtual environment is of central importance in VR. In this context, different kinds of tracking systems using a diversity of mediums (e.g., optical, magnetic, ultrasonic, and inertial) enable the position and orientation of human users and physical objects to be calculated within a physical space in real time. This is particularly important when calculating the correct viewing perspective for the user [18]. Coined with gesture recognition software, tracking systems allow natural body movements to be translated into functional interaction techniques [101] . Handheld controllers allow users to navigate and manipulate objects in the virtual world [23]. To improve interactions, haptic devices provide force feedback through physical manipulators, allowing a better understanding of how objects in a virtual environment interact with each other [82]. Other kinds of feedback such as vibration, wind, temperature, and pressure, can also be integrated in the virtual environment given the proper feedback devices [18]. In one simple word, VR is all about illusion [63], where user experience is of central importance. VR solutions should convince users that they feel physically located within the virtual world [18]. Generating a sense of presence sets VR apart from other digitalization technologies [22]. As the renowned illusionist Harry Houdini stated: "What the eyes see and the ears hear, the mind believes."

VR has been increasingly used to improve various activities in smart industries, and has introduced the concept of virtual manufacturing (VM). VM is a VR solution that allows generating information about the structure, status, and behavior of a manufacturing system similarly to a real manufacturing environment [69] . It offers a modeling and simulation platform to simulate the production of products and the building of assembly lines, along with their associated manufacturing processes [63]. For instance, BMW Group uses the NVIDIA Omniverse platform to simulate its entire production process with photo-realistic details using its SORDI dataset, including physical properties like gravity and different materials [78]. The VR environment is coupled with different artificial intelligence (AI) tools, including a battery of supervised, unsupervised, and reinforcement learning algorithms to control robots and industrial machines, and simulate human workers' behavior on the virtual factory floor. NVIDIA CEO Jensen Huang recurrently discussed BMW's use of Omniverse during his keynotes at the company's annual GTC conference in April 2021–2023: "NVIDIA, initially known for its gaming platforms and graphics chips, has broadened its scope to training AI-based programs for industrial applications including automotive manufacturing and medical imaging" [78].

3.6.2   Augmented Reality

Augmented reality (AR) is a variation of VR where virtual objects are overlaid on the real environment [9], allowing virtual and real objects to coexist together and interact with each other in real-time in the real environment [26]. While VR creates a virtual environment and attempts to trick human users' senses to experience it in a realistic way, differently, AR integrates virtual objects into a real three dimensional environment to augment it while allowing a seamless experience for the users. To make sure that the overlapping of virtual objects is consistent, AR systems estimate the virtual objects' orientation and position in real time. This can be done in different ways, such as using specified markers in the real environment, identified by cameras and matched against predefined patterns [75]. More recently, markerless techniques have been introduced, namely Natural Feature Tracking (NFT) and Simultaneous Localization And Mapping (SLAM). NFT-based solutions utilize computer vision models to detect certain representative points representing natural features in the real-time video images [48]. Visual feature tracking algorithms are then used to produce accurate motion estimates and compute the virtual objects' pose accordingly [48]. SLAM-based solutions consist in building a probabilistic feature map of the real environment, in real-time, in the form of a 3D point cloud, and then determines the AR navigation paths as a pre-scanning process. The 3D point cloud live tracking has improved over time, especially after commercial sensors became capable of detecting environment structure information, using technologies such as structured light, wireless dosimeters, and X-ray propagation [20, 89]. 3D objects constructed based on data from the sensors are compared with the predefined virtual models to estimate the virtual objects' poses [11].

AR solutions can be categorized according the devices being used. For instance, head mounted displays including googles and helmets can be organized in two groups: *optical see-through* and *video see-through* [9]. With optical see-through solutions, a half-transparent mirror permits users to see the real world and reflect virtual information into the user's eyes, combining the real and virtual objects accordingly [109]. With video see-through solutions, the real world is digitized previously, allowing the digital models to be merged with the real environment before being shown as an opaque display [10]. Another category of AR solutions uses mobile screens, such as smart phones and tables, to display virtual objects, where a coupled camera captures the real environment while a dedicated computer device renders the virtual image and projects in onto the screen [109]. Some solutions have introduced spatial augmented reality (SAR), using projectors to project the virtual information directly onto the real objects, simplifying user interaction with the virtual object [10]. AR can also be augmented with VR functionality, making the virtual assets more tangible so that they can interact with the user. The latter is usually referred to as mixed reality (MR), compared with typical AR which mainly focuses on projecting the virtual elements into the real environment. MR solutions allow virtual objects to be rotated, scaled, or explored in different ways. This is especially useful in logistics for instance, where a packaging planner can place virtual components in a real pallet or container, using finger gestures to adjust their size and try positioning them in different directions [115].

From an industrial perspective, AR promotes an improved collaboration among designers and technicians, and enhances the handling and availability of data [25]. Its industrial applications range over manual product assembly, robot programming and operations, maintenance, process monitoring, training, process simulation, quality inspection, picking process, operational setup ergonomics and safety [26]. AR solutions can provide work instructions to employees on how to execute a certain activity. Instructions can be generated automatically by a procedural or machine learning software agent, or they can be provided by a remote human expert – reducing the need for on-site expert presence [26]. AR can also improve employees' safety on the factory floor, especially when involving human-robot interactions [26], by emphasizing robot movements and danger zones, and delineating the safe zone as a virtual overlay on top of the real video images.

3.7 Robotics

Robotics includes the design, building, maneuver, and usage of robots [85]. It is an interdisciplinary branch of computer science and computer, electrical, and mechanical engineering that develops machines to assist or replace humans, and imitate human behavior [121]. Various robotics solutions are deployed and utilized today in different contexts and scenarios, namely in smart industries and manufacturing processes, where they are proving essential in performing delicate tasks (e.g., fabricating, building, cutting, and painting material with extreme precision) and operating in dangerous environments (e.g., dealing with hazardous industrial materials, working in high heat, underwater, or in space). Robots come in different sizes, shapes, and forms, from humanoids who are made to resemble humans in appearance and simulate human behavior, to industrial robots and cobots. Industrial robots typically appear in the form of robotic arms or grippers and are mainly involved in the machining and assembly of products, hidden behind safety barriers to protect employees [115]. In contract, collaborative robots or cobots are design to work with humans and complement them in the production or logistics process. Cobots are designed for safety: they are generally made of flexible material and are usually smaller in size compared with their heavy weight industrial counterparts (cf. Fig. 3.10).

Fig. 3.10 Industrial robotic arm in BMW Group' US plant in Spartanburg, South Carolina [21] (**a**) and (**b**) Collaborative robot (cobot) in BMW Group's Leipzig plant, Germany [44]

Human-machine collaboration through software agents and cobots is at the center of the Industry 5.0, where machines help humans extend their craftsmanship and analytical skills to deliver higher quality products and services. Allowing seamless collaboration between humans and machines is not an easy task, and requires a huge amount of planning, training, and testing, in order to achieve a useful and safe collaboration [115]. Autonomous Mobile Robots (AMR) are another king of robots consisting of a driverless transport of materials and goods within warehouses and factory floors. They come in different shapes and sizes, from robots that replace the forklift truck and move pallets independently, to pulling trailers with a tractor (cf. Fig. 3.5). They move around using a battery of cameras and sensors for navigation and obstacle avoidance. More recent systems like the idealworks iw.hub produce complete 3D scans of their environment creating their own virtual navigation maps accordingly. Iw.hubs utilize NVIDIA Isaac SDK and the Robot Operating System (ROS) – an open-source robotics middleware suite for robot software development. Initially developed at the AI laboratory at Stanford [3], ROS provides a set of tools for low-level device control, messaging between processes, and integration with outside libraries to allow real-time computing, navigation, and decision making. Another example of a robotics programming platform is NVIDIA's Isaac SDK, which is designed for optimal performance on NVIDIA's Jetson boards and GPUs. Frameworks like ROS and Isaac SDK, coined with modular designs, and simpler programming and visual simulations tools, are paving the way toward a large-scale adoption of robotics systems in industry, transforming factories into automated cyber-physical systems.

3.8   Cyber-Physical Systems

The new generation of smart manufacturing systems consist of the collaborative integration of humans, physical systems, and cyber systems [84]. Humans are the creators, managers, and users of the physical system, the physical system is the factory or industry providing the manufacturing process, and the cyber system is the software layer allowing to analyze, calculate, and control the manufacturing process [41]. The term Cyber-Physical System (CPS) was first introduced by the US National Science Foundation (NSF) around 2006, describing a smart physical system controlled and monitored by an intelligent cyber system [154]. Hardware and software components in a CPS are genuinely intertwined, and interact with each other at different spatial and temporal scales, with different behavioral modalities, according to their context [47]. A CPS is technically designed as a network of interacting robotic and sensor elements with physical input and output instead of a stand-alone device [71]. A nice example of such a system is the Distributed Robot Garden at MIT[5] where a group of robots tend a garden of plants, combining distributed sensing (each plant is equipped with a sensor monitoring its status), navigation, manipulation and wireless networking [37]. Other examples include smart grid, autonomous automobile systems, industrial control systems, and medical monitoring systems, among others [122]. Note that the integration of hardware and software components in manufacturing is not new. The term Computer Integrated Manufacturing Systems (CIMS) has been used to describe the integrated manufacturing system that combines physical processes with computing [154]. Nonetheless, early CIMSs adopted centralized control schemes limited inside of the factory floor, and suffered from the lack context-awareness, flexibility, and self-configuration [154]. New digitalization technologies like the Internet of Things (IoT), sensor networks, and Radio-Frequency Identification (RFID), coined with Artificial Intelligence (AI) and Machine Learning (ML) algorithms, provide advanced and unprecedented monitoring and control capabilities of real-word processes, which were not available with legacy CIMSs.

A CPS connects the physical world with the cyber space using a communication network (cf. Fig. 3.11). The physical world refers to the physical objects, processes, or the environment to be monitored or controlled (such as a factory plant, or a supply chain). The cyber space refers to the next generation information infrastructure, including services, applications, and decision-making software agents, which we refer to as the Intelligent Web (described in the following Sect. 2.11). The communication network refers to intermediate network components which allow communication and information exchange between the physical world and its cyber space counterpart. The Internet provides many mature network technologies such as IP/TCP, XML/JSON, access control, network link, publish/subscribe model, among others. However, the realization of full-fledged CPSs requires new technologies such as mobile node localization, semantic analysis of heterogeneous data, sensor network coverage, and mass data transmission, which can be provided through the well anticipated Intelligent Web (i.e., the software layer of the IoT, cf. Sect. 2.11).

CPSs represent a central opportunity area and a major source of competitive advantage for the innovation economy in the twenty-first century [154], and have attracted wide attention from the industry, academia, and governments [83]. The United States' NSF has provided enormous funds to promote research and innovation on CPS, highlighting its huge potential impact on US national interests [92].

---

[5] Massachusetts Institute of Technology.

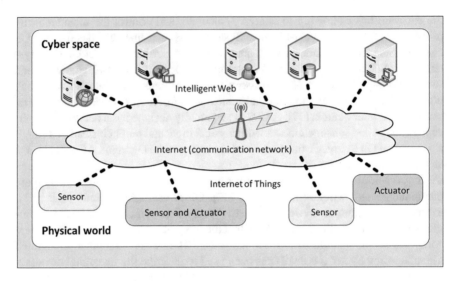

Fig. 3.11 Abstract overview of a CPS [154]

The European Commission has also promoted CPS-related research in its Horizon 2020–30 programs [113]. The United States' President's Council of Advisors on Science and Technology reports recommend CPS as one of the six transformative civil technologies driving American economic growth, and consider it as a core opportunity area and source of competitive advantage for the US [29]. We will further discuss industry empowered CPSs and their usage in smart manufacturing in Chap. 3 of this book.

3.9   Evolution of the Web Toward Collective Knowledge and Intelligence

Starting in the 1990s, the digitization of business records coincided with the commercial success of the Web. This presented a major change in how data records are represented in the early twenty-first century. In the early 1990s, many Internet-based projects were in development. These include HTML (Hypertext Markup Language) which codes the graphical design and contents of Web pages (HTML documents), where the content is accessed through a so-called "browser": a general purpose viewer of Web pages (e.g., Google Chrome, Microsoft Edge). Developments also include HTTP (Hypertext Transfer Protocol): a protocol to retrieve and modify Web pages, as well as URIs (Universal Resource Identifiers): providing a universal addressing scheme used to locate Web pages. In this context, it is important to distinguish between two concepts which are often confused together: the Internet and Web. On the one hand, the Internet is a network of networks originally developed as a US

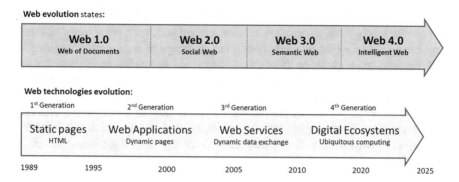

Fig. 3.12 Evolution of the Web

military project in the 1970s, extended to general public use since the mid-1980s, and consisting of a set of protocols and standards managing computer connectivity such as TCP (Transport Control Protocol), IP (Internet Protocol), etc. On the other hand, the Word Wide Web is a software that allows information sharing across a network: namely the Internet. It provides the software platform including a user-friendly interface to access data shared on the Internet, based on the concepts of hypertext (hypermedia) and URIs linking information with hyperlinks, such as with HTML document links where information is uniquely identified using URIs.

3.9.1 From Static to Intelligent Web

The Web has known a fast evolution: going from the Web 1.0, known as Web of Documents where users are merely consumers of static information, to the more dynamic Web 2.0, known as social or collaborative Web where users produce and consume information simultaneously, and entering the more sophisticated Web 3.0, known as the Semantic Web by giving information a well-defined meaning so that it becomes more easily accessible by human users and automated processes. Fostering service intelligence and atomicity (the ability of autonomous services to interact automatically), remains one of the most upcoming challenges of the Semantic Web. This promotes the dawn of a new era: the Intelligent Web (or Web 4.0), known at the physical layer as the Internet of Things (IoT), an extension of the Semantic Web where (physical/software) objects and services autonomously interact in a multimedia virtual environment, provided with embedded communication capabilities, common semantics and addressing schemes, promoting the concept of Digital Web Ecosystems where everywhere (human and software) agents collaborate, interact, compete, and evolve autonomously in order to automatically solve complex and dynamic problems (cf. Fig. 3.12). The Intelligent Web extends the concept of cooperation from *User-User* with the Social Web (2.0) and *User-Machine* with the Semantic Web (3.0) to *Machine-Machine* interaction, where machines do not simply interact with humans but rather interact with each other based not only on a simple exchange of data but rather on an exchange of semantically meaningful information, i.e., knowledge that is well understood by machines. It promotes the vision of digital ecosystems where human and software agents collaborate, interact, compete, and evolve autonomously in order to automatically solve complex and dynamic problems. While we are currently in the middle of the Semantic Web era, yet we are steadily moving toward the Intelligent Web era, especially in relatively controlled environments like smart manufacturing and digitalized industrial applications.

3.9.2   From Big Data to Collective Knowledge

The Semantic Web vision aims at associating machine-readable semantic descriptions to Web data, using two major technological breakthroughs: (i) knowledge bases (such as taxonomies and/or ontologies [15, 118]), which provide predefined semantic information references (similarly to dictionaries for human users) to allow the identification and extraction of semantic meaning from raw data, and (ii) dedicated data representation technologies (namely RDF/S for linked data description [77], OWL for ontology definition [35], and SPARQL for semantic data manipulation and querying [116]). These technologies are extensible, interoperable, and platform-independent [36], aiming to improve data modeling, annotation, manipulation, search and integration, and thus allowing intelligent information retrieval on the Web, which is at the core of the Semantic Web. Yet, developing service intelligence and atomicity, i.e., the ability of software agents and services to interact and sustain themselves automatically, without human interaction, remains one of the most upcoming challenges of the Semantic Web. In addition, Semantic Web technologies and social networking services are promoting a new form of collaboration: nowadays, it is common for Web users to contribute their multimedia data and knowledge to the community, allowing the editing and manipulation of such public knowledge in a collaborative environment (e.g., Wikis, blogs, Foursquare,[6] Google Latitude,[7] etc.). As a result, the Web is becoming more than a distributed container of (raw and/or semantic) multimedia data, but is increasingly harnessing *collective knowledge*, viewed as the combination of all known data, information, and meta-data concerning a given (set of) concept(s), fact(s), user(s), or processes (s), as well as the semantic links between them [7]. Hence, software agents (and/or intelligent terminals) are expected to automatically analyze and handle large collections of multimedia data with their contents,

---

[6] Location-based social website for mobile devices (http://Foursquare.com).

[7] Location-aware mobile application allowing users to view their contacts geographic locations (www.google.com/latitude). Note that Google Latitude is being recently retired, transforming most of its services to Google+.

links and transactions, using the sum of their respective intelligence and knowledge, in order to improve data accessibility, management, and exchange between people and computers. Also, agents (terminals) in the Intelligent Web vision are expected to guaranty autonomous data/services sustainability and evolution, e.g., predicting future events that might affect the data/services, and acting therefore in order to preserve or update them accordingly.

Collective knowledge management has been addressed from different perspectives in different application scenarios, including Web-related and smart industry applications. For example, collaborative social tagging of Web resources is viewed as an attempt of acquisition and sharing of so-called collective knowledge concerning a given community. Wikis have also become popular tools for collaboration on the Web among many vibrant online communities promoting CK extraction and representation. Hence, extracting and handling collective knowledge requires intelligent Web terminals (agents) which are not only capable of understanding and meaningfully processing information (using SW technologies), but which are also capable of thoroughly collaborating and even "reasoning" together, as a collective, to produce and handle common collective knowledge, leading to more sophisticated (intelligent) services, as well as achieving the ultimate goal: *collective intelligence*, where Web agents are able to automatically sustain themselves and evolve without direct human intervention.

3.9.3   Toward Collective Intelligence

An Intelligent Web (IoT) terminal is viewed as a cyber-physical entity having a software agent capable of handling IoT technologies in order to enhance collaboration between humans and machines. Yet, individual intelligence needs to be coordinated in order to enhance its own capabilities as well as the capabilities of its surrounding entities. In this perspective, collective intelligence will emerge, which consists in effectively mobilizing the skills of a group of agents in a digital ecosystem to emerge and handle the collective knowledge from all agents. The new form of collective knowledge will be automatically aggregated and recombined to create new knowledge, new rules, and/or new ways of learning what individual agents of the ecosystem cannot do individually. For instance, knowledge recommendation methods can be developed to identify from the large pool of maintained collective knowledge, the pieces of knowledge and data contents which are required by an agent (human user or automated process), based on explicit needs, past experiences, profile, and preferences. In addition to knowledge exchange and manipulation, as Intelligent Web (IoT) terminals move and interact within their environment, events will be automatically generated (e.g., service requested from provider, action performed by client, at a certain location, etc.). These events need to be subsequently enhanced with relevant knowledge in order to describe the context in which each event happened (such as why a thing was observed at a location, or how and why it interacted with another thing) and act accordingly. This highlights the need for innovation to automatically interpret events and processes related to Web terminals (agents) in given contexts, adding semantic annotations and predicting what will happen, and what precautionary measures could be taken to optimize data/services sustainability and evolution. Such issues can be handled through the use of artificial intelligence and machine learning, as digital ecosystems will likely solve problems by evolving solutions, e.g., starting with a current set of semantically-rich events (i.e., a set of *solutions*), and then iteratively learning from them through supervised, unsupervised, and reinforcement learning processes to produce enhanced and more useful solutions over time.

3.10   IoT Technologies and Semantic Interoperability

3.10.1   Internet of Things

The Internet of Things (IoT) underlines the concept of a network of networks, linking public/private infrastructures, dynamically extended by connection points consisting of the "things" (terminals) connecting to one another. Enhanced processing capabilities and always-on connectivity, will make terminals gain a central role in communications: terminals (deemed henceforth intelligent) will be able to form bridges between existing network structures thus extending the overall infrastructure capacity [138]. In this context, developments in network technologies such as RFID (Radio Frequency Identification), short-range wireless technologies, and sensor networks, coupled with enhancements in network addressing techniques, such as IPv6 to expand address space, become critical to the IoT, allowing to connect more objects in the physical/virtual worlds. Yet, scalability and cross platform compatibility between varied networked systems remain an open problem. Network technologies need to offer solutions that allow ubiquitous access, i.e., connecting any terminal to the network, which will require dedicated network protocol translation gateways (defining the correspondences between varied network and communication protocols), compared with today's IP (Internet Protocol) which only allows end-to-end communication between devices without any midway protocol translation. Here, improvements in wireless and sensor communication protocols (from direct transmission and minimum transmission energy [143], to multi-hop routing, multi-path routing, and cluster-based routing [150]) can be utilized to improve scalability and robustness for dynamic networks, reducing the amount of information that must be transmitted between terminals (e.g., integrating data fusion within the routing protocol [45]) in order to improve connectivity. Also, safeguarding effective and ubiquitous connectivity for terminals requires extensive terminal design efforts covering: (i) *mobility*, allowing occasional or continuous mobility of terminals in the selected environment, (ii) *resources,* and *energy efficiency*, since the terminals' resources availability might vary from limited (e.g., with sensor terminals) to unlimited (e.g., with Cloud computing systems), (iii) supporting various *communication modalities*, ranging over electromagnetic communication (radio frequency), to optical, acoustic, as well as inductive and capacitive coupled communications, and (iv) supporting various *network topologies*, such as single hop, star, multi-hop, mesh and/or multitier. Yet, the sheer diversity of terminals which will be supported stipulates that no single hardware or software platform can hope to support the whole design space [61]. Thus, heterogeneous systems and interoperability need to be addressed.

3.10.2   Semantic Interoperability

In this context, semantic interoperability becomes a vital requirement: allowing smart terminals and sensors to communicate "meaningfully" with each other, exchanging data (knowledge) and processes (services) despite the heterogeneous nature of the underlying information structures and communication protocols. Luckily, the issue of semantic interoperability has been at the center stage of Semantic Web (SW)-related studies, introducing standard data representation and manipulation technologies (e.g., XML, RDF, OWL, and SPARQL) to simplify information and knowledge interchange. It has also been investigated in related domains, namely Service Oriented Architectures (SOA), aiming to improve communication and information exchange between heterogeneous service providers and requestors [105]. Service requestors/providers in SOA are generally dynamic, operating on the "publish-find-bind" paradigm principle, where services are dynamically added and described (published) in a service registry. The service descriptions are then used to search (find) and associate (bind) the service to the service requestor. The problem of semantic interoperability is more severe in such dynamic situations due to the lack of predefined relationships between the requestors/providers [61]. On one hand, the development of shared information models using SW technologies (e.g., shared RDF or OWL reference ontologies defining common semantics following the SW vision) can improve semantic interoperability among the participant terminals and sensors [79]. Nevertheless, the problems with this approach remain: (i) the complexity of producing a universal ontology (encompassing all semantic descriptions concerning all possible terminals and processes), and (ii) semantic rigidity, underscoring the difficulty in updating or extending the reference ontology once it is defined (in order to handle new terminals, new processes, and new information [51]). We note that there is an ongoing work currently conducted by W3C Semantic Sensor Network Incubator Group aiming to provide an ontological representation of a sensor network to order to solve these problems.[8] On the other hand, semantic interoperability can be achieved by producing appropriate semantic mediators (translators) at each terminal's end, to allow the conversion to the information (knowledge) format which the terminal understands. A combination of context independent shared information models can be utilized (using

---

[8] http://www.w3.org/2005/Incubator/ssn/XGR-ssn-20110628/; https://www.w3.org/TR/vocab-ssn-ext/

SW data representation technologies such as RDF and OWL), coined with context specific information specialization approaches (using SW data manipulation technologies such as SPARQL) to achieve semantic interoperability [162]. This semantic mediation approach avoids imposing a unique information model (e.g., a unique reference ontology) that has to be agreed and adopted by everybody, thus allowing intelligent Web terminals (agents) to select the formats best suited for their needs, while still being able to interact and use services offered by other terminals [162]. Yet, defining semantic mediators for each IoT terminal and smart sensor does not seem feasible in practice and remains an open issue. Hence one can predict that a certain compromise between shared semantic references and semantic mediators can be made.

3.11 Cloud and Edge Computing

Cloud computing is a dynamic computing model where a large pool of systems and machines are connected through a network to provide a scalable infrastructure for applications, computation, content storage and delivery [73]. It allows on-demand provisioning of (potentially unlimited) computational resources as a utility, where computing resources are offered as a metered service similar to a physical public utility like electricity, water, natural gas, or the telephone network (cf. Fig. 3.13a). Cloud computing enables a computing system to acquire or release computing resources on demand in a manner that is (virtually) transparent to the user. It relies on the underlying concepts of *utility computing*: the combination of computing resources as a metered service in a way similar to a physical public utility, *scalability*: the ability of a computing system to grow relatively easily in response to increased demand, *elasticity*: The ability of a system to dynamically acquire or release compute resources on-demand, and *high availability*: systems designed such that the loss of any one component of a system will not result in system failure [65].

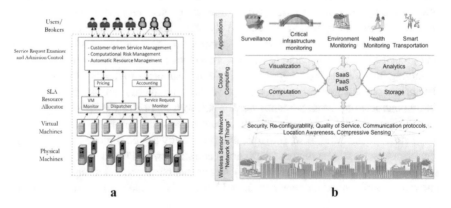

Fig. 3.13 Cloud computing overall architecture (**a**) and integration with IoT (**b**). (**a**) Cloud computing overall architecture. (**b**) Conceptual IoT framework with cloud computing at the center [59]

A consumer can acquire services from: (i) a full computer infrastructure (raw processing capabilities) – i.e., Infrastructure as a Service (IaaS), (ii) a software platform (operating and/or development systems) – i.e., Platform as a Service (PaaS), or (iii) turnkey applications (ready to use software solutions) – i.e., Software as a Service (SaaS) [74]. IaaS allows the provisioning of resources necessary to build an application environment from scratch, e.g., servers (computing capacity), connections (networking capacity), and memory (storage capacity), to run, maintain and distribute Information Technology (IT) resources more efficiently and cost-effectively compared with traditional IT infrastructure delivery models. Sample IaaS solutions include Amazon Web Services (AWS), VMWare, 3Tera, and XCalibre. While IaaS solutions allow a high degree of infrastructure flexibility, yet they are usually complex to work with, and likely require some specialized expertise to make them work [73]. PaaS for application deployment on top of an IaaS cloud allows the building and delivery of web applications and services including application design, application development, testing and hosting, database management, application versioning, and multi-tenant architectures, among others. Some PaaS examples include Google Cloud (Python), Heroku (Ruby and Rails), AWS, Elastic, Beanstalk, and Microsoft Azure Pipelines. PaaS allow straightforward application development and deployment with little effort, and a large degree of scalability as the applications are cloud-based, and they do not require for systems administrators as they are part of the service itself. Nonetheless, PaaS services might come with restrictions or trade-offs especially with a pre-existing application to be ported to the PaaS: the application might require special libraries, tools, operating system requirements, etc., which are not available on the PaaS of choice, requiring to code around these issues to deploy on the PaaS which is not always optimal [65]. SaaS provides a software delivery model via application hosting on the Cloud. It pre-dates the cloud with Web application hosting solutions. In that regard, SaaS is rather concerned with server hosting management, allowing a higher degree of Quality of Service (QoS) control and management. Famous SaaS solutions include Google Workspace, Salesforce, Pipedrive, and Microsoft Office 365. Note that SaaS is not the ultimate goal of cloud computing which focuses primarily on IaaS and PaaS models, yet it provides an important and useful functionality [74].

More recently, cloud computing has shown great promise in establishing flexible IoT platforms, allowing to receive data from a network of pervasive sensors, analyzing and processing the data starting from the edge nodes, moving in toward the inner nodes of the cloud, and providing easy to understand Web based visualizations, where the sensing and processing work in the background hidden from the user. Here, we can emphasize edge computing as a disturbed computing model that brings data processing closer to the data sources, i.e., performing part of the data processing (including data cleaning, filtering, and sometimes data transformation), before transmitting the data to the inner nodes of the cloud. Edge computing aims at improving response time and reducing bandwidth consumption over the network, while exploiting the available resources at the edge nodes. Edge computing is usually regarded as a specific cloud computing architecture or a network topology emphasizing location-sensitive distributing computing [97].

Cloud computing can serve as middleware between IoT and sensor network infrastructures on the one hand, and software applications on the other hand (cf. Fig. 3.13b). Sensing service providers can join the network and offer their data using a storage cloud, analytic tool developers can provide their software tools, artificial intelligence experts can provide their data mining and machine learning tools useful in transforming information to knowledge, and computer graphics designers can offer a variety of visualization tools [59]. Cloud computing can offer these services as Infrastructures, Platforms, or Software, where the data generated, knowledge created, tools used, and the product visualizations disappear into the cloud, thus allowing to tap the full potential of the IoT in many application domains [88] . The IoT application specific cloud framework should be able to provide support for (i) reading data streams either from sensors directly or fetch the data from databases, (ii) easy expression of data analysis logic as functions/ operators that process data streams in a transparent and scalable manner on Cloud infrastructures, and (iii) if any events of interest are detected, outcomes should be passed to output streams which are connected to a visualization program [59]. Using such frameworks, the developers of IoT applications are able to utilize cloud computing services without getting bugged down with the individual creation, handling, and scalability of individual applications [88].

3.12   Big Data and Data Analytics

Big data underlines massive and complex data sets, of different data-types and data formats, generated through the use of different devices, and managed by different organizations and companies through advanced architectural solutions [65]. Big data is often characterized by the so-called 5Vs model: Volume, Velocity, Variety, Veracity, and Value. *Volume* refers to the huge amount of data generated from different sources, ranging over manufacturing, financial services, healthcare, and social media and entertainment, among others. The International Data Corporation (IDC) reports that manufacturing had the largest share of data in 2018 (i.e., 3585 Exabyte) and will have a 30% annual growth rate between 2018 and 2025 (the second highest growth rate behind healthcare data's 36% growth) [117]. *Variety* refers to the different types of data collected via sensors, smartphones, or social networks. Data types include video, image, text, audio, and data logs, in either structured, unstructured, or semi-structured format. Note that most industrial data recordings in recent years comes in semi-structured format like JSON, XML, image, video, and audio metadata descriptions, while fewer data is reported in structured tabular forms like relational databases or spreadsheets [149]. *Velocity* underlines the speed of data transfer, resulting from the introduction of new data collections, re-usage of archived data, and the arrival of streamed data from multiple sources [19]. *Veracity* underlines the quality of the data that is collected, which depends on certain factors like the reliability and quality of the source the data is collected from, the quality of the processes following which the data is collected, and the way the data will be utilized [32]. The veracity of a user's data underlines how reliable and significant the data really is for the user [90]. *Value* underlines the process of discovering value form the data, in the form of extracted insights, semantics, knowledge, and analytics [30]. Issues of the 5Vs become more intertwined as more industries generate big data, where the same or different types of data come from the same or different devices with various sampling frequencies, formats, and precision levels, which makes it challenging to extract value-added insights [32]. A larger number of big data tools have been developed for industry, including Apache Hadoop-based sensor data management framework for could manufacturing [12], Cloud-Based Design Manufacturing (CBDM) model [152], and fog-computing based framework to monitor machine health in cyber-manufacturing [106]. The authors in [32] highlight some of the differences between handling big data in manufacturing and handling big data on the Web. Industrial data is produced and supported by various industrial vendors and associations such as manufacturers of actuators, meters, sensors, controllers, and software companies. Manufacturers use different hardware interfaces, communication protocols, manufacturing processes, machine-readable languages, and semantic definitions [32]. In contrast, most data on the Web is based on natural languages, is represented using some Web standards or standard-like representations, and is thus easier to be exchanged without the difficulties associated with multiple interfaces and protocols. Moreover, manufacturing systems have well defined objectives, intricate and sophisticated functions, and are meticulously specialized for specific application scenarios. In comparison, social media data for instance is much looser in its definition, aim, and scope, where big data tools collect and store time series data from millions of customers who follow each other online to report trending events. In industrial applications, time series data is collected from multiple sources, integrated, and analyzed to explain and predict the specific states of manufacturing entities [32]. For instance, the sensor data collected from an industrial auto-part gripper machine is used to reflect the state of the machine and can be developed to produce a virtual simulation model of the machine, predicting its behavior and performing preventive maintenance.

Industrial *big data analytics* refers to analyzing and interpreting big industrial data, potentially in near-real-time, in order to drive intelligent insights and initiate closed-loop corrective actions [132]. Recent data analytics tools are mostly AI-powered using a battery of machine learning algorithms, designed to find patterns in huge time series for industrial use cases such as predictive maintenance, real-time quality control, and scenario testing for root cause analysis [132]. For instance, the streaming computation engine Spark is becoming a popular tool compare with its traditional Hadoop MapReduce batching engine counterpart [32]. In fact, batch processing is not able to provide real-time analytics response such as real time monitoring, dynamic scheduling and planning on systems of workshop floor (such as with Supervisory Control And Data Acquisition –SCADA, systems). In this context, Microbatching and streaming engines (like Spark, Storm, Flink) can provide real-time big data analytics of streaming data [106]. Spark makes use of batching and streaming to replace the MapReduce engine. Furthermore, compared with Storm and Flink, Spark has more powerful analytics tools such as SparkSQL, Spark R, GraphX, and MLlib [32]. However, Storm and Flink typically outperform Spark in real-time processing performance, and are usually utilized together with Spark for more powerful computation functionality [32]. Apache Beam provides a uniform abstraction layer to run these real-time engines together at the execution layer [56]. While the latter engines are still at their early usage stages in industry, further research needs to be done in order to better identify the big data issues that are of specific importance in manufacturing, compared with the capabilities of existing big data tools, and the main components required to design new and more sophisticated big data solutions for industry.

3.13   Digital Twin and Digital Thread

A digital twin is a virtual and real-time resemblance of a physical process, product, or service occurring in the real world [64]. It can be viewed as a digital mirror, sibling, or simply a twin of a physical system [83]. The concept of digital twin initially came from the aerospace field to analyze and predict the behavior and performance of aircraft systems, namely NASA (US National Aeronautics and Space Agency) spacecraft and space shuttles [54]. It can represent the performance, operations, environment, geometry, and resource states of a system or product based on the continuously collected data and real-time updates from its real-world counterpart. A digital twin allows real-time interaction and convergence between the physical space and the cyber space [135]. This is especially useful in manufacturing where digital twins are increasingly used to model digital factories that mirror their real-world counterparts (such as the BMW Regensburg factory, cf. Fig. 3.5). A digital twin can be decoupled into three main parts: (i) the 3-D model, (ii) the mathematical model, and (iii) the rule model. The 3-D model should look exactly like the physical system, and is produced using 3D scanning and graphics rendering techniques discussed previously in this chapter. In an industrial setting, the 3D model often represents a Virtual Factory Layout (VFL) which allows to replicate a physical factory's geometry in the virtual environment. The mathematical model allows representing the mechanisms, kinetics, and laws of physics (gravity, humidity material erosion, etc.) related to the physical system, and is usually run on powerful simulation engines like Unity 3D and NVIDIA Omniverse (e.g., the latter is currently adopted by BMW Group in developing its Regensburg plant and the related digital twin projects). Several methods can be used for mathematical modeling, which we broadly categorize as knowledge-driven and data-driven. On the one hand, knowledge-driven solutions use kinetics and data with some system assumptions represented by human experts or software agents in well-defined knowledge bases or ontologies. Proper inference mechanisms are defined to extract the needed knowledge and utilize it in describing the system's behavior. On the other hand, data-driven solutions use Artificial Intelligence (AI) methods to learn and consequently model the systems' laws and characteristics. Dedicated Machine Learning (ML) algorithms are used to mine the available data from the physical system in order to learn the needed rules and patterns to build its digital counterpart [83]. The rule model defines the interactions from users and allows to represent and control the behavior of the digital twin in coordination with its physical counterpart. This is achieved through the IoT and sensor network infrastructure that is set-up in the real-world, using it to connect with the Cyber-Physical System (CPS) that governs the functioning of the physical system, in order to enable a full digitalization and virtualization of the physical system and its environment.

As modern products become more complex and increasingly personalized, companies are struggling to respond in a timely fashion to the market demand, albeit with assured quality and competitive costs. In this context, USAF Global Science and Technology Vision Task Force recently introduced the concept of *digital thread*, as a data-driven architecture that links together information generated across the product lifecycle, providing a data exchange and communication platform for a company's products at any point in time during the product's lifecycle [144]. In other words, the digital thread handles the massive data flow that will be generated from the production line, ranging over the processes of data collection, storage, aggregation, analysis, and exchange, in order to provide timely information to manufacturers and clients [32]. The digital thread can help reduce holding and operational cost on machines, transport, and warehousing by providing timely information on idle and faulty equipment. It can also help identify potential production problems and improve production efficiency by selecting appropriate maintenance time schedules [134]. The digital thread provides data that can be shared among manufacturers, suppliers, and customers, allowing to identify the best suppliers and top clients, while providing flexibility in adapting product design and manufacturing according to available supplies and customer preferences [102]. The latter requires real-time interaction between manufacturers, suppliers, and clients, which can be provided through an IoT enabled Cyber-Physical System (CPS) industry model, or preferably through a fully digitized digital twin model. Note that while digital threads can be implemented with legacy CPSs [130], nonetheless, their impact and potential is most perceived when integrated with digital twins [32].

3.14   Toward the Industrial Metaverse

There has been increasing commotion and excitement recently about the industrial metaverse, with initiatives like BMW's fully digitized Regensburg plant before building the physical facility [50], Mercedes Benz's MO360 Data Platform connecting its car manufacturing plants to the Microsoft Cloud to improve predictability across its digital production and supply chain [28], and Boeing's digital twin development model to design its airplanes – allowing to achieve up to 40% improvement in first-time quality of the parts and systems it uses to manufacture commercial and military airplanes [17]. According to Boeing CEO Dennis Muilenburg, "this model is going to be the biggest driver of production efficiency improvements for the world's largest airplane maker over the next decade" [17].

***

But what is the industrial metaverse exactly? Is it another fancy name for digital twin? Or is it a digital twin empowered with a digital thread?

***

While digital twins and digital threads are central to this new paradigm, yet the digital metaverse has "a much larger scale with increasing complexity by creating digital twins of entire systems such as factories, airports, cargo terminals, or cities—not just digital twins of individual machines or devices that we have seen so far", as expressed by Thierry Klein, president of Bell Labs at Nokia [119]. Also, the industrial metaverse promises an immersive experience, using Virtual Reality (VR) and Augmented Reality (AR) to provide immersive user interaction, environment sensing, and operational feedback, allowing users to switch between the physical factory represented as a Cyber-Physical System (CPS) and the virtual factory represented as a digital twin, in a seamless way as they are identical replicas of each other. While the digital twin provides a virtual mirror of the physical system, using VR and AR technologies allow for an improved and immersive experience within the digital twin and outside of it, allowing to blur the boundaries between the real physical processes and their digital counterparts.

In other words, the industrial metaverse can be seen as a culmination of the integration of multiple maturing digitalization technologies, ranging over digital twins, IoT, VR, AR, and AI, which aims to fuel the growth of industry and the optimization of its operations [119]. More specifically, the industrial metaverse paradigm promises to enhance visibility, flexibility, planning, and risk management, which would improve production line and supply chain resiliency and risk mitigation [127]. This will help manufacturers innovate in more efficient and streamlined ways, by virtually designing, developing, and testing industrial processes and remote solutions. Optimizing industrial designs through the metaverse paradigm would also help promote sustainable development, by dabbling with and integrating innovative processes to reduce emissions and natural resource consumption and incorporate sustainable materials into products, while meeting the customers' needs [127]. The industrial metaverse can be realized by integrating digital Information Technology (IT) with physical Operation Technology (OT), to make real-time data-driven decisions and create new products and services. It will integrate IoT-based sensor streams of real time OT data for continuous analysis to monitor production quality, process performance, equipment effectiveness, and quality of service. The data can be stored and analyzed through dedicated cloud computing platforms to extract historical patterns and identify future trends, providing the needed analytics and insights and combining them with the operational digital twin in order to simulate potential scenarios, identify possible future opportunities, and prevent potential setbacks. Digital twins will provide different ways to visualize and interact with the data, products, and processes, and their associated real-time telemetry [127]. Certain digital twins can focus on emphasizing the relationship between IT and OT to improve efficiency and resiliency, while other twins can be designed to build and test new products, studying their performance and simulating their impacts using dedicated machine learning algorithms (e.g., encoder-decoder models, sequence-to-sequence models), before they are pushed into physical production [28] . AI-tools can also be used to automate and improve part of the interaction with clients and employees, using advance Natural Language Processing (NLP) models and chat bots to assist users in daily routine tasks. Coupled with VR and AR technology, the metaverse will provide its users with an immersive collaboration experience helping both customers and frontier workers improve their work effectiveness, efficiently, safety, and overall experience [129].

By the end of this decade, the industrial metaverse is expected to be a multitrillion-dollar market [129]. More importantly, it can be one of the greatest forces to drive both sustainability and the digital transformation of businesses and industries. It will make innovation easier, progress quicker, time to market lesser, while reducing waste and natural resource consumption [129]. It also promises to develop improved and more personalized products by allowing clients and workers to explore more alternative designs in shorter time and at significantly lower costs. It can also allow an easier integration of recycling and circular economy principles into the design process, investigating more efficient ways of production [129]. Nonetheless, this next stage of digitalization will also be a challenge for most companies. Yet, this is an inevitable challenge that will need to be addressed sooner rather than later, providing an opportunity to build and cultivate a virtual world that will help solve real-world problems. As mentioned in our first chapter, history taught us there is no turning back on technology. This is especially true with the industrial metaverse movement that is about to transform all of our industries. Manufacturers will have to choose: either lead the way, or be dragged behind and get swept away by the industrial metaverse current…

References

1. R. Abboud, J. Tekli, *MUSE Prototype for Music Sentiment Expression.* IEEE International Conference on Cognitive Computing (ICCC'18), part of the IEEE World Congress on Services 2018, 2018. pp. 106–109
2. R. Abboud, J. Tekli, Integration of non-parametric fuzzy classification with an evolutionary-developmental framework to perform music sentiment-based analysis and composition. Soft Comput. **24**(13), 9875–9925 (2019)
3. E. Ackerman, E. Guizzo, *Wizards of ROS: Willow Garage and the Making of the Robot Operating System.* IEEE Spectrum: Technology, Engineering, and Science News, 2017. https://spectrum.ieee.org/wizards-of-ros-willow-garage-and-the-making-of-the-robot-operating-system
4. R. Al Sobbahi, J. Tekli, Comparing deep learning models for low-light natural scene image enhancement and their impact on object detection and classification: overview, empirical evaluation, and challenges. Signal Process. Image Commun. **109**, 116848 (2022)
5. S.R. Al, J. Tekli, Low-light homomorphic filtering network for integrating image enhancement and classification. Signal Process. Image Commun. **100**, 116527 (2022)
6. E. Alpaydin, *Introduction to Machine Learning*, 4th edn. (MIT, 2020) pp. xix, 1–3, 13–18, ISBN 978-0262043793

7. R. Armbrecht et al., Knowledge management in research and development. Res. Technol. Manag. **44**(4), 28–48(21) (2001)
8. J. Attieh, J. Tekli, Supervised term-category feature weighting for improved text classification. Knowl. Based Syst. **261**, 110215 (2023)
9. R. Azuma, A survey of augmented reality. Presence Teleop. Virt. **6**(4), 355–385 (1997)
10. R. Azuma et al., Recent advances in augmented reality. IEEE Comput. Graph. Appl. **21**, 1–27 (2001)
11. H. Bae et al., Fast and scalable structure-from-motion based localization for high-precision mobile augmented reality systems. J. Mob. User Exp. **5**, 4 (2016)
12. Y. Bao et al., *Massive Sensor Data Management Framework in Cloud Manufacturing Based on Hadoop*. EEE International Conference on Industrial Informatics (INDIN'12), 2012. pp. 397–401
13. L. Barghout, Visual taxometric approach to image segmentation using fuzzy-spatial taxon cut yields contextually relevant regions, in *Information Processing and Management of Uncertainty in Knowledge-Based Systems*, (Springer, 2014)
14. D. Batista et al., *Semi-Supervised Bootstrapping of Relationship Extractors with Distributional Semantics*. Conference on Empirical Methods in Natural Language Processing (EMNLP), 2015. pp. 499–504
15. M. Baziz et al., *A Concept-Based Approach for Indexing Documents in IR*. INFORSID 2005, 2005. pp. 489–504, Grenoble, France
16. B. Becerik-Gerber et al., Assessment of target types and layouts in 3D laser scalllling for registration accuracy. Autom. Constr. **20**(5), 649–058 (2011)
17. W. Bellamy, *Boeing CEO Talks 'Digital Twin' Era of Aviation* (Avionics International, 2018) https://www.aviationtoday.com/2018/09/14/boeing-ceo-talks-digital-twin-era-aviation/
18. L. Berg, J. Vance, Industry use of virtual reality in product design and manufacturing: a survey. Virtual Reality **21**, 1–17 (2017)
19. J.J. Berman, *Principles of Big Data: Preparing, Sharing, and Analyzing Complex Information*. Springer, eBook. ISBN: 9780124047242, 2013
20. M. Billinghurst et al., A survey of augmented reality. Found. Trends Human Comput. Interact. **8**, 73–272 (2015)
21. BMW Group, *Innovative Human-robot cooperation in BMW group production*. Press release (2013) https://www.press.bmwgroup.com/global/article/detail/T0209722EN/innovative-human-robot-cooperation-in-bmw-group-production?language=en
22. D. Bowman, R. McMahan, Virtual reality: how much immersion is enough? Comput. Graphics Forum **40**(7), 36 (2007)
23. D. Bowman et al., 3D user interfaces: new directions and perspectives. IEEE Comput. Graph Appl. **28**(6), 20 (2008)
24. S. Brewster, A. Gies, *The Best VR Headset*. New York Times, 2023. https://www.nytimes.com/wirecutter/reviews/best-standalone-vr-headset/
25. F. Bruno et al., Visualization of industrial engineering data in augmented reality. J. Vis. **9**(3), 319–329 (2006)
26. L. Cardoso et al., A survey of industrial augmented reality. Comput. Ind. Eng. **139**, 106159 (2020)
27. J. Carew, *Reinforcement Learning*. TechTarget. Accessed June 2023. https://www.techtarget.com/searchenterpriseai/definition/reinforcement-learning#:~:text=Reinforcement%20learning%20is%20a%20machine,learn%20through%20trial%20and%20error
28. M.N. Center, *Mercedes-Benz and Microsoft Collaborate to Boost Efficiency, Resilience and Sustainability in Car Production*. news.microsoft.com, 2022. https://newsmicrosoft-com/2022/10/12/mercedes-benz-and-microsoft-collaborate-to-boost-efficiency-resilience-and-sustainability-in-car-production/
29. U.D.T.I. Center, *Disruptive Civil Technologies: Six Technologies With Potential Impacts on US Interests Out to 2025*. 2008. https://apps.dtic.mil/sti/citations/ADA519715
30. M. Chen et al., Big data: a survey. Mob. Netw. Appl. **19**(2), 1–39 (2014)

31. C. Cruz-Neira et al., *Surround-screen projection-based virtual reality: the design and implementation of the CAVE.* Proceedings of the 20th annual conference on Computer graphics and interactive techniques, 1993. pp 135–142

32. Y. Cuia et al., Manufacturing Big Data ecosystem: a systematic literature review. Robot. Comput. Integr. Manuf. **62**, 101861 (2020)

33. M. Dasso, T. Constant, M. Fournier, The use of terrestrial LiDAR technology in forest science: application fields, benefits and challenges. Ann. For. Sci. **68**(5), 959–974 (2011)

34. R. Davies, *Machine Vision: Theory, Algorithms, Practicalities.* Morgan Kaufmann, 2005. ISBN 978-0-12-206093-9

35. M. Dean, G. Schreiber, *OWL Web Ontology Language Reference.* W3C Recommendation, http://www.w3.org/TR/owl-ref/. 2004

36. S. Decker et al., The semantic web: the roles of XML and RDF. IEEE Internet Comput. **4**(5), 63–74 (2000)

37. J. DelPretro, D. Rus, *Distributed Robot Garden.* MIT-Computer Science & Artificial Intelligence Laboratory, 2020. https://www.csail.mit.edu/research/distributed-robot-garden

38. B. El Asmar et al., *AWARE: A Situational Awareness Framework for Facilitating Adaptive Behavior of Autonomous Vehicles in Manufacturing.* International Semantic Web Conference (ISWC'20), 2020. (2): 651–666

39. A. Eriksson et al., *Virtual Factory Layouts from 3D Laser Scanning – A Novel Framework to Define Solid Model Requirements.* 7th CIRP Conference on Assembly Technologies and Systems 76:36–41

40. M. Evans, *From Nepal to Idaho, Inter Breaks Groung in Virtual Reality* (Idaho National Laboratory, 2019) https://inl.gov/article/from-nepal-to-idaho-intern-breaks-ground-in-virtual-reality/

41. Y. Fan et al., A digital-twin visualized architecture for flexible manufacturing system. J. Manuf. Syst. **60**, 176–201 (2021)

42. M. Fares et al., Unsupervised word-level affect analysis and propagation in a lexical knowledge graph. Knowl.-Based Syst. **165**, 432–459 (2019) Elsevier

43. M. Fares et al., *Difficulties and Improvements to Graph-based Lexical Sentiment Analysis using LISA.* IEEE International Conference on Cognitive Computing (ICCC'19), 2019. pp. 28–35

44. M. Farish, *A Collaborative Approach to Automation.* Automotive Manufactoring Solutions (AMS) (2020) https://www.automotivemanufacturingsolutions.com/technology/a--collaborative-approach-to-automation/41400.article

45. C.H. Feng et al., UPS: unified protocol stack for emerging wireless networks. Ad Hoc Networks Special Issue on Cross-layer Design in Ad Hoc and Sensor Networks **11**, 687–700 (2013) Elsevier

46. S. Ferilli et al., *Towards Sentiment and Emotion Analysis of User Feedback for Digital Libraries.* Italian Research Conference on Digital Libraries (IRCDL'16), 2016. pp. 137–149

47. U.N.S Foundation, *Cyber-Physical Systems (CPS).* 2010. https://www.nsf.gov/pubs/2010/nsf10515/nsf10515.htm

48. P. Fraga-Lamas et al., *A Review on Industrial Augmented Reality Systems for the Industry 4.0 Shipyard.* IEEE Access, 2018. 13358–13375

49. V. Francisco et al., Ontological reasoning for improving the treatment of emotions in text. Knowl. Inf. Syst. **25**(3), 421–443 (2010)

50. J. Friedrich, *All BMW Group Vehicle Plants to be Digitalised Using 3D Laser Scanning by Early 2023.* BMW Group Press Club, 2022. https://www.press.bmwgroup.com/global/article/detail/T0400833EN/all-bmw-group-vehicle-plants-to-be-digitalised-using-3d-laser-scanning-by-early-2023?language=en

51. R. Garcia-Castro, A. Gomez-Perez, Interoperability results for semantic web technologies using OWL as the interchange language. J. Web Semant. **8**(4), 278–291 (2010)

52. M.F. Gavilanes et al., Creating emoji lexica from unsupervised sentiment analysis of their descriptions. Expert Syst. Appl. **103**, 74–91 (2018)

53. M. Ghiassi, S. Lee, A domain transferable lexicon set for twitter sentiment analysis using a supervised machine learning approach. Expert Syst. Appl. **106**, 197–216 (2018)
54. E. Glaessgen, D. Stargel, *The Digital Twin Paradigm for Future NASA and U.S. Air Force Vehicles.* 53rd AIAA/ASME/ASCE/AHS/ASC Structures, Structural Dynamics and Materials Conference, 2012. https://ntrs.nasa.gov/citations/20120008178
55. A. Glassner, *Principles of Digital Image Synthesis*, 2nd edn. (Kaufmann, San Francisco, 2004) ISBN 978-1-55860-276-2
56. M. Gokalp et al., *Big-Data Data Analytics Architecture for Businesses: A Comprehensive Review on New Open-Source Big-Data Tools* (Cambridge Service Alliance, 2017), pp. 1–35
57. Y. Goldberg, A primer on neural network models for natural language processing. J. Artif. Intell. Res. **57**, 345–420 (2016)
58. U. Govindarajan et al., Immersive technology for human-centric cyberphysical systems in complex manufacturing processes: a comprehensive overview of the global patent profile using collective intelligence. Complexity **2018**, 17 (2018)
59. J. Gubbi et al., Internet of Things (IoT): a vision, architectural elements, and future directions. Futur. Gener. Comput. Syst. **29**(7), 1645–1660 (2013)
60. S. Guha et al., *Clustering Data Streams.* Proceedings of the Annual Symposium on Foundations of Computer Science (FOCS), 2000. pp. 359–366
61. P. Guillemin, P. Friess, *The Internet of Things: Strategic Research Agenda.* CERP-IoT – Cluster of European Research Projects on the Internet of Things, 2010. Vision and Challenges for Realizing the Internet of Things, Ch 3, pp. 41–42
62. A. Hajjar, J. Tekli, *Unsupervised Extractive Text Summarization Using Frequency-Based Sentence Clustering.* European Conference on Advances in Databases and Information Systems (ADBIS'22), 2022. pp. 245–255
63. N. Hamid et al., Virtual reality applications in manufacturing system. Sci. Inf. Conf., 1034–1037 (2014)
64. H. Harb, H. Noueihed, *Digital Twin's Promising Future in Digital Transformation* (JOUN Technologies, 2020) 15 p
65. I. Hashem et al., The rise of "big data" on cloud computing: review and open research issues. Inf. Syst. **47**, 98–115 (2015)
66. K. Hille, *NASA Turns to AI to Design Mission Hardware.* NASA Space Tech, 2023. https://www.nasa.gov/feature/goddard/2023/nasa-turns-to-ai-to-design-mission-hardware Accessed March 2023
67. J. Hoffart et al., YAGO2: a spatially and temporally enhanced knowledge base from Wikipedia. Artif. Intell. **194**, 28–61 (2013)
68. S. Hussain, M. Haris, A K-means based co-clustering (kCC) algorithm for sparse, high-dimensional data. Expert Syst. Appl. **118**, 20–34 (2019)
69. K. Iwata et al., Virtual manufacturing systems as advanced information infrastructure for integrated manufacturing resources and activities. CIRP Ann. **46**, 335–338 (1997)
70. A. Junyi et al., *SpeechT5: Unified-Modal Encoder-Decoder Pre-training for Spoken Language Processing.* Annual Meeting of the Association for Computational Linguistics (ACL), 2022. (1), pp. 5723–5738
71. H. Kang et al., Smart manufacturing: past research, present findings, and future directions. Int. J. Precis. Eng. Manuf. Green Technol.. 3:(1)111–128
72. M. Kearns et al., A sparse sampling algorithm for near-optimal planning in large Markov decision processes. Mach. Learn. **49**(193–208), 193–208 (2002). https://doi.org/10.1023/A:1017932429737
73. A. Khajeh-Hosseini et al., Research challenges for Enterprise cloud computing. CoRR abs/1001.3257, 2010
74. A. Khajeh-Hosseini et al., The cloud adoption toolkit: supporting cloud adoption decisions in the enterprise. Softw. Pract. Exper. **42**(4), 447–465 (2012)
75. D. Khan et al., Factors affecting the design and tracking of ARToolKit markers. Comput. Stand. Interfaces **41**, 56–66 (2015)

76. L. Klein et al., Imaged-based verification of Asbuilt documentation of operational building. Autom. Constr. **21**(I), 161–171 (2012)
77. G. Klyne, J. Carroll, *Resource Description Framework (RDF): Concepts and Abstract Syntax*. W3C Recommendation REC-rdf-concepts-20040210, 2004. http://www.w3.org/TR/rdf-concepts/
78. W. Knight, *BMW's Virtual Factory Uses AI to Hone the Assembly Line*. Wired, 2021. https://www.wired.com/story/bmw-virtual-factory-ai-hone-assembly-line/
79. J. Krogstie et al., Integrating semantic web Technology, web services, and workflow modeling: achieving system and business interoperability. Int. J. Enterp. Inf. Syst. **3**(1), 22–41 (2007)
80. K. Kumar et al., A hybrid deep CNN-Cov-19-res-net transfer learning architype for an enhanced brain tumor detection and classification scheme in medical image processing. Biomed. Signal Process. Control **76**, 103631 (2022)
81. J. Lai et al., Semi-supervised feature selection via adaptive structure learning and constrained graph learning. Knowl. Based Syst. **251**, 109243 (2022)
82. S. Laycock, A. Day, A survey of haptic rendering techniques. Comput. Graph. Forum. **26**(1), 50 (2007)
83. Z. Lei et al., Toward a web-based digital twin thermal power plant. IEEE Trans. Industr. Inform. **18**(3), 1716–1725 (2022)
84. J. Leng et al., Digital twin-driven joint optimisation of packing and storage assignment in large-scale automated high-rise warehouse product-service system. Int. J. Comput. Integr. Manuf. **1–18** (2019)
85. G.N. Library, *International classification system of the German National Library (GND)*. Accessed March 2023. https://portal.dnb.de/opac.htm?method=simpleSearch&cqlMode=true&query=nid%3D4261462-4
86. E. Lindskog et al., Production system redesign using realistic visualisation. Int. J. Prod. Res., 2016. 55(3): 858–869 (2017)
87. Z. Liu et al., Joint video object discovery and segmentation by coupled dynamic Markov networks. IEEE Trans. Image Process **27**(12), 5840–5853 (2018)
88. T. Lopez et al., Adding sense to the internet of things an architecture framework for smart objective systems. Pers. Ubiquit. Comput. **16**, 291–308 (2012)
89. R.N. Loy, N. Padoy, Seeing is believing: increasing intraoperative awareness to scattered radiation in interventional procedures by combining augmented reality, Monte Carlo simulations and wireless dosimeters. Int. J. Comput. Assist. Radiol. Surg. **10**, 1181–1191 (2015)
90. T. Lukoianova, Veracity roadmap: is big data objective, truthful and credible? Adv. Classif. Res. Online **24**(1), 4–15 (2014). https://doi.org/10.7152/acro.v24i1.14671
91. Y. Ma et al., Background augmentation generative adversarial networks (BAGANs): effective data generation based on GAN-augmented 3D synthesizing. Symmetry **10**(12), 734 (2018)
92. J. Marburger et al., *Leadership Under Challenge: Information Technology R&D in a Competitive World. An Assessment of the Federal Networking and Information Technology R&D Program*. US Defence Technical Information Center, 2007. https://apps.dtic.mil/sti/citations/ADA474709
93. S. Marschner, *Monte Carlo Ray Tracing*. Cornell University Computer Science CS4620, 2013
94. MathWorks, *What Is Deep Learning? 3 Things You Need to Know*. Accessed June 2023. https://www.mathworks.com/discovery/deep-learning.html#:~:text=Deep%20learning%20is%20a%20machine,a%20pedestrian%20from%20a%20lamppost
95. H. Maziad et al., *Preprocessing Techniques for End-to-End Trainable RNN-Based Conversational System*. International Conference on Web Engineering (ICWE), 2021. pp. 255–270
96. S. Mehta et al., *Towards Semi-Supervised Learning for Deep Semantic Role Labeling*. Conference on Empirical Methods in Natural Language Processing (EMNLP), 2018. pp. 4958–4963
97. M. Merenda, C. Porcaro, D. Iero, Edge machine learning for AI-enabled IoT devices: a review. Sensors **20**(9), 2533 (2020)

98. G.A. Miller, C. Fellbaum, WordNet then and now. Lang. Resour. Eval. **41**(2), 209–214 (2007)
99. S. Mishra, J. Diesner, *Semi-Supervised Named Entity Recognition in Noisy-Text*. International Conference on Computational Linguistics (COLING), 2016. pp. 203–212
100. T. Mitchell, *Machine Learning* (McGraw Hill. ISBN 0-07-042807-7. OCLC 36417892, New York, 1997)
101. S. Mitra, T. Acharya, Gesture recognition: a survey. IEEE Trans. Syst. Man Cybern. C **37**(3), 311 (2007). https://doi.org/10.1109/TSMCC.2007.893280
102. W. Mohammed et al., *Configuring and visualizing the data resources in a cloud-based data collection framework*. International Conference on Engineering, Technology and Innovation (ICE/ITMC'17), 2017. pp. 1201–1208
103. M. Mohri et al., *Foundations of Machine Learning* (The MIT Press, 2012) https://mitpress. mit.edu/9780262039406/foundations-of-machine-learning/
104. T. Morris, *Computer Vision and Image Processing* (Palgrave Macmillan, 2004) ISBN 978-0-333-99451-1
105. M. Nagarajan et al., *Semantic Interoperability of Web Services – Challenges and Experiences*. Proceedings of the Fourth IEEE International Conference on Web Services (ICWS'06), 2006. pp. 373–382
106. K. Nagorny et al., Big Data analysis in smart manufacturing: a review. Int. J. Commun. Netw. Syst. Sci. **2017**(10), 31–58 (2017)
107. NASA, *The Virtual interface Environment Workstation (VIEW)*. National Aeronautics and Space Administration, 1990. https://www.nasa.gov/ames/spinoff/new_continent_of_ideas/
108. S. News, *Climate Change: Seven Technology Solutions that Could Help Solve Crisis*. 2021. https://news.sky.com/story/climate-change-seven-technology-solutions-that-could-help-solve-crisis-12056397
109. A. Nishihara, *Object Recognition in Assembly Assisted by Augmented Reality System Object Recognition in Assembly Assisted by Augmented Reality System*. SAI Intelligent Systems Conference (IntelliSys), 2015. https://doi.org/10.1109/IntelliSys.2015.7361172
110. H. Noueihed et al., *Simulating Weather Events on a Real-World Map Using Unity 3D*. Proceedings of the International Conference on Smart Cities and Green ICT Systems (SMARTGREENS'22), 2022. pp. 86–93
111. H. Noueihed et al., Knowledge-based virtual outdoor weather event simulator using Unity 3D. J. Supercomput. **78**(8), 10620–10655 (2022)
112. T. Oates, D. Jensen, *The Effects of Training Set Size on Decision Tree Complexity*. International Conference on Machine Learning (ICML'97), 1997. pp. 254–262
113. R. Owen et al., Responsible research and innovation: from science in society to science for society with society. Sci. Public Policy **39**(6), 751–760 (2012)
114. M. Pharr, G. Humphreys, *Physically Based Rendering from Theory to Implementation* (Elsevier/Morgan Kaufmann, Amsterdam, 2004) ISBN 978-0-12-553180-1
115. A. Pinker, M. Pruglmeier, *Innovations in Logistics*. Huss, 2021. 192 p
116. E. Prudhommeaux, A. Seaborne, *SPARQL Query Language for RDF*. W3C Recommendation, 2008. http://www.w3.org/TR/rdf-sparql-query/
117. D. Reinsel et al., *Data Age 2025: The Digitization of the World from Edge to Core*. https://www.seagate.com/files/www-content/ourstory/trends/files/idc-seagate-dataage-whitepaper.pdf (2018)
118. P. Resnik, Using information content to evaluate semantic similarity in a taxonomy. Proc. Int. Joint Conf. Artif. Intell. **1**, 448–453 (1995)
119. MIT Technology Review, *The Industrial Metaverse – A Game-Changer for Operational Technology*. 2023. https://www.technologyreview.com/2022/12/05/1063828/the-industrial-metaverse-a-game-changer-for-operational-technology/
120. C. Rooney, R. Ruddle, *HiReD: A High-Resolution Multi-Window Visualisation Environment for Cluster-Driven Displays*. ACM SIGCHI Symposium on Engineering Interactive Computing System (EICS'15), 2015. pp. 2–11
121. S. Russel, P. Norvig, *Artificial Intelligence, A Modern Approach,* 3rd, Pearson, 2015. 1164 p

122. S. Khaitan, J. McCalley, Design techniques and applications of cyberphysical systems: a survey. IEEE Syst. J. **9**, 2 (2014)

123. K. Salameh et al., *SVG-to-RDF Image Semantization*. 7th International SISAP Conference, 2014. pp. 214–228

124. K. Salameh et al., Unsupervised knowledge representation of panoramic dental X-ray images using SVG image-and-object clustering. Multimedia Syst. (2023). https://doi.org/10.1007/s00530-023-01099-6

125. G. Salloum, J. Tekli, Automated and personalized nutrition health assessment, recommendation, and progress evaluation using fuzzy reasoning. Int. J. Human-Comput. Stud. **151**, 102610 (2021)

126. G. Salloum, T. Tekli, Automated and personalized meal plan generation and relevance scoring using a multi-factor adaptation of the transportation problem. Soft Comput. **26**(5), 2561–2585 (2022)

127. C. Sanders, *Industrial Metaverse: The Data Driven Future of Industries.* Microsoft Industry Blogs, 2023. https://www.microsoft.com/en-us/industry/blog/manufacturing/2023/02/13/industrial-metaverse-the-data-driven-future-of-industries/#:~:text=The%20industrial%20metaverse%20is%20redefining,improvements%20in%20sustainability%20and%20efficiency

128. Y. Shoham et al., *Multi-agent Reinforcement Learning: A Critical Survey.* Technical Report, Stanford Universitt, 2003. pp. 1–13

129. Siemens, *What Is the Industrial Metaverse – And Why Should I Care?* Siemenscom, 2023. https://www.siemens.com/global/en/company/insights/what-is-the-industrial-metaverse-and-why-should-i-care.html

130. V. Singh, K. Willcox, *Engineering Design with Digital Thread.* MIT Libraries, DSpace@MIT, 2021. https://dspace.mit.edu/handle/1721.1/114857

131. M. Sonka et al., *Image Processing, Analysis, and Machine Vision* (Thomson. ISBN 978-0-495-08252-1, 2008)

132. B. Stackpole, D. Greenfield, *Big Data.* Automation World, 2022. https://www.automationworld.com/analytics/article/22485289/big-data

133. F.G. Taddesse et al., *Semantic-Based Merging of RSS Items.* World Wide Web J. Internet Web Inf. Syst. J. Spec Issue Human-Centered Web Sci 2010. 13(1–2): 169–207, Springer

134. F. Tao et al., Manufacturing service management in cloud manufacturing: overview and future research directions. J. Manuf. Sci. Eng **137**(2015), 040912 (2015)

135. F. Tao et al., Digital twin-driven product design, manufacturing and service with big data. Int. J. Adv. Manuf. Technol. **94**, 3563–3576 (2018)

136. O. Taylor, A. Rodriguez, Optimal shape and motion planning for dynamic planar manipulation. Auton. Robot. **43**(2), 327–344 (2019)

137. J. Tekli et al., Minimizing user effort in XML grammar matching. Inf. Sci. **210**, 1–40 (2012) Elsevier

138. J. Tekli et al., *Semantic to Intelligent Web Era: Building Blocks, Applications, and Current Trends.* International Conference on Managment of Emergent Digital EcoSystems (MEDES), 2013. pp. 159–168

139. J. Tekli, An overview on XML semantic disambiguation from unstructured text to semi-structured data: background, applications, and ongoing challenges. IEEE Trans. Knowl Data Eng. **28**(6), 1383–1407 (2016)

140. J. Tekli et al., Full-fledged semantic indexing and querying model designed for seamless integration in legacy RDBMS. Data Knowl. Eng. **117**, 133–173 (2018)

141. J. Tekli, An overview of cluster-based image search result organization: background, techniques, and ongoing challenges. Knowl. Inf. Syst. **64**(3), 589–642 (2022)

142. A. Tewari et al., State of the art on neural rendering. Comput. Graphics Forum **39**(2), 701–727 (2020)

143. S. Tilak et al., A taxonomy of wireless micro-sensor network models. ACM Mob. Comput. Commun. Rev. **6**(2), 28 (2002)

144. USAF Global Science and Technology Vision, T.F., *Global Horizons Final Report*. Homeland Security Digital Library, 2021. https://www.hsdl.org/c/
145. A. Valdivia et al., Sentiment analysis in TripAdvisor. IEEE Intell. Syst. **32**(4), 72–77 (2017)
146. A. Valitutti et al., Developing affective lexical resources. PsychNology J. **2**(1), 61–83 (2004)
147. D. Vilares et al., Universal, unsupervised (rule-based), uncovered sentiment analysis. Knowl.-Based Syst. **118**, 45–55 (2017)
148. O. Vinyals, Q. Le, *A Neural Conversational Model*. CoRR abs/1506.05869, 2015
149. S. Wang et al., Knowledge reasoning with semantic data for real-time data processing in smart factory. Sensors **18**, 1–10 (2018)
150. T. Wang et al., *Link Energy Minimization for Wireless Sensor Networks*. Elsevier Ad Hoc Networks Special Issue on Cross-layer Design in Ad Hoc and Sensor Networks, 2012. 10(3):569–585
151. T. Warren, *A Closer Look at HTC's New Higher-Resolution Vive Pro*. The Verge, 2018. https://www.theverge.com/2018/1/9/16866240/htc-vive-pro-vr-headset-hands-on-ces-2018
152. D. Wu, D. Rosen, et al., Cloud-based design and manufacturing: a new paradigm in digital manufacturing and design innovation. Comput. Aided Des. **59**, 1–14 (2015). https://doi.org/10.1016/j.cad.2014.07.006
153. Q. Xie et al., *Unsupervised Data Augmentation for Consistency Training*. Conference on Neural Information Processing Systems (NeurIPS), 2020
154. X. Yao et al., Smart manufacturing based on cyber-physical systems and beyond. J. Intell. Manuf. **30**(8), 2805–2817 (2019)
155. D. Yaworsky, *Word-Sense Disambiguation Using Statistical Models of Roget's Categories Trained on Large Corpora*. Proceedings of the International Conference on Computational Linguistics (Coling), 1992, vol 2, pp. 454–460. Nantes
156. D. Zacharopoulou et al., *A Web-based Application to Support the Interaction of Spatial and Semantic Representation of Knowledge*. AGILE: GIScience Series, 2022. 3:70
157. S. Zhang et al., Sentiment analysis of Chinese micro-blog text based on extended sentiment dictionary. Future Gener. Comput. Syst. **81**, 395–403 (2018)
158. T. Zhang et al., *BIRCH: An Efficient Data Clustering Method for Very Large Databases*. Proceedings of the ACM SIGMOD Conference on Management of Data, 1996. 25(2):103–114
159. T. Zhang et al., Fairness in graph-based semi-supervised learning. Knowl. Inf. Syst. 2023. 65(2): 543–570 (2023)
160. Z. Zhang et al., *Moving Object Recognition for Airport Ground Surveillance Network*. International Conference on Mobile Networks and Management (MONAMI'21) 2021. pp. 335–343
161. F. Zhou et al., A survey of visualization for smart manufacturing. J. Vis. **22**, 419–435 (2019)
162. W. Zhu, S. Vij, *Extending SOA Infrastucture for Semantic Interoperability*. 3rd Annual DoDSOA & Semantic Technology Symposium, 2011. Alion Science and Technology

Chapter 4
How Visual Data Is Revolutionizing the Industry World

Gathering industrial data and applying it to advance products, processes, and services is nothing new. Customer surveys, sales reports, query logs, and other types of data have long been used to pinpoint problems and inform business strategies accordingly [1]. Images and video data have also been used throughout the manufacturing process, from product design, production, testing, and deployment, to customer feedback. Legacy Computer Aided Design (CAD) tools help designers create their product designs, refine their appearance, and provide the needed design schematics to launch them into production. Monitoring sensors and cameras provide scalar and visual feedback on the production, testing, and deployment phases of the products, allowing to identify product defects, verify machining processes, and check equipment installations [2]. Nonetheless, the problem of legacy systems predating the Industry 4.0 is they were often slow to provide useful insights, and the sample data sizes that were collected were often two small to ensure data accuracy and infer useful insights [1]. Also, early solutions dealing with image and video data focused on two dimensional data, and were limited by computer processing capabilities and visualization methods [2].

Nowadays, the amount of the multimedia data generated by digitalized businesses is unparalleled and is increasingly reaching record highs, as more sensors and IoT (Internet of Things) terminals are deployed in the workplace, as more powerful GPU-enabled (Graphical Processing Unit) computer systems and Cloud computing services become available to store and process large quantities of multimedia data, as recent AR (Augmenter Reality) and VR (Virtual Reality) solutions allow immersive three dimensional visualizations of products and processes, and as digital twins create virtual models of complete manufacturing plants blurring the boundaries between the virtual model and its real counterpart. It is those organizations that are embracing digitalization that stand a chance to gain a competitive edge [1].

Industry 4.0 technologies have placed digital multimedia data at the center of organizations of all sizes and across many industries. Nonetheless, harnessing the power of digitalization goes beyond investing in the right technologies and equipment. It is not enough to set-up a camera system, deploy an IoT infrastructure, purchase Cloud resources, or develop a digital twin of the factory floor. Businesses without a complete data strategy will have a lot of difficulty making sense of the huge influx of multimedia data, and risk falling behind their better-prepared competitors.

***

How is digitalization transforming industrial applications and methods? How are visual and multimedia data leveraged to improve industrial processes?

***

We attempt to answer these questions in this chapter and in the remainder of this book…

4.1 Industrial Production Process

Industry 4.0 digitalization technologies allow production machinery and equipment to generate massive quantities of data on a continuing basis, with the purposes of creating new values for both manufacturer and customers [3]. Industrial data is usually characterized by its multimedia nature, including signal sensing data, tabular and text data, and image and video data. The data needs to be processed automatically, providing users with interactive ways to facilitate useful pattern identification and valuable information exploration. This enables human-in-the-loop decision-making through the different phases of the industrial process, ranging over design, production, testing, training, and service provision.

4.1.1 Design Phase

The design phase is a creative thinking process during which designers create and refine the appearance, function, and performance of a product based on market demand [4]. Modern design tasks are becoming increasingly more intricate, including different kinds of standards and criteria that modern products need to comply with, and more demanding user requirements that products and services need to fulfil. In this context, digitalization can help alleviate the complexity of the design task by enabling so-called data-driven CAD (Computer Aided Design) [5]: providing the designer with design-related data to guide the design process through computer software. The digital representation of the product in the virtual world will show the designers' expectations and simulate the designs' physical constraints from the physical world. This will help designers cross-examine their designs' constraints in both virtual and physical worlds and adjust them accordingly [6]. Design-related data helps explore the constraint associations between different criteria (be it performance-, security-, reliability-, availability-, usability-, visual appeal- or esthetics-related, among others), while verifying the functional requirements of the design prototype [7]. Yet design constraints do not always match, and are often times contradicting. For instance, designing the most visually appealing sports car does not always result in the most practical designs, or the best performing designs, since some compromise might be made on behalf of practicality and performance, for the sake of the car chassis' visual appeal. To help alleviate this problem, recent multi-view CAD solutions have been recently utilized in various industries, e.g., [8–10], to allow considering multiple criteria and providing multifaceted visualizations, including product, worker, and production line perspectives (cf. Fig. 4.1). Combined with automatic evaluation algorithms, modern CAD solutions allow designers to interactively and iteratively change and adjust the proposed product design, in the hopes of fulfilling the required design criteria.

Recently, the introduction and increased affordability of 3D printing has changed CAD solutions from producing effective designs to performing rapid prototyping. 3D printing allows converting the digital model generated by CAD software into solid samples that can be studied physically, and can even be directly applied to product manufacturing [11]. This can significantly bring down costs and can simplify the transition to the production phase [4].

4.1.2   Production Phase

The production phase transforms product designs from conceptual models and their visualizations, into physical implementations. Production usually occupies the largest share of production costs in the manufacturing life cycle [4]. It is mainly concerned with the formulation and management of the production process, aiming to maximize production efficiency and minimize production costs. Yet achieving the latter goals is easier said than done, especially with the intricacies of production processes. In this context, digitalized and interconnected modern production lines in the form of Cyber Physical Systems (CPSs) and more recently digital twins, provide a vast amount of data that helps alleviate the complexity of the production phase while optimizing its performance and cost. Digitalized production line provides real-time data to monitor the production process and track the status of the produced product (cf. Fig. 4.2). They allow real-time interaction with the production workers for onsite troubleshooting, and provide production managers with valuable insights through the recording and post processing of historical production data which can be data-mined offline using dedicated Machine Learning (ML) and data analytics tools for process improvement and innovation.

On the one hand, real-time production data analysis allow production workers to observe and monitor the production process as it unfolds, using live data captured from the production line. This allows evaluating the operation status of the production line components and the status of the products being manufactured, in order to handle abnormal situations in a timely way [14]. The real-time monitoring of

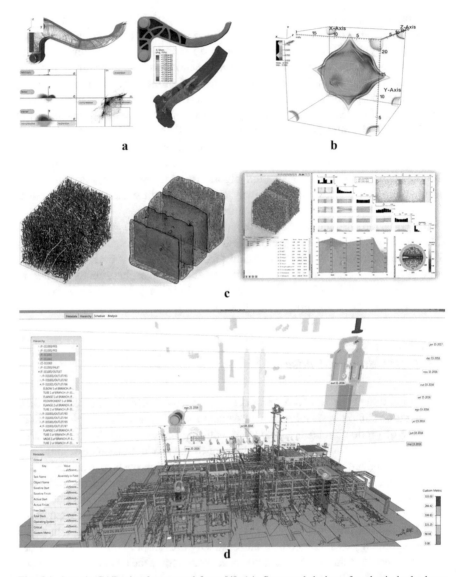

Fig. 4.1 Sample CAD visuals reported from [4]. (**a**). Structural design of a plastic brake lever product comparing the mechanical stress performances of different brake lever designs [9]. (**b**). Material characteristics visualizations [10]. (**c**). Helping designers understand the structural characteristics of different materials to select the one most fitting the product requirements [10]. (**d**). Production environment design allowing planners to identify schedule uncertainties and work–space conflicts in the virtual layout [8]

production lines is performed through a continuous data stream produced by the IoT infrastructure that is built within the factory's CPS. On the other hand, historical data analysis aims at extracting hidden patterns within the recorded production data, and discovering useful insights from historical data trends, in order to facilitate

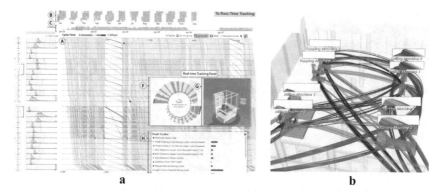

Fig. 4.2 Sample visual monitoring tools reported from [4]. (**a**). Assembly line performance monitoring and troubleshooting [12]. (**b**). CPS bottleneck exploration allowing user-guided examination of bottleneck steps and overloaded devices [13]

production process optimization and management innovation [4]. With the help of advanced visual analytics, data is provided to the process managers, providing interactive functions to allow human-computer collaborative intelligent analysis of the data [15] (e.g., filtering, transforming, zooming, clustering, classifying, and extrapolating the data to acquire useful insights).

4.1.3   Testing Phase

Testing is the process that measures and evaluates the functions and performance levels of products and processes [16]. It aims to assure that the products are manufactured according to established standards, while fulfilling the user requirements and preferences. Testing usually relies on well-defined protocols as well as lessons learned from previous tests to guide the evaluation process. The testing phase is arguably the most data hungry phase in the entire manufacturing process, since it involves the processing and crunching of huge amounts of data according to the adopted test protocol and experiments. Also, testing often requires multiple sets of experimental runs with different input parameters to verify the range of product parameters and their suitability, generating multidimensional and multivalued testing data results which are not easy to process by humans [17]. In this context, AI-enabled data analytics and visualization tools like Tableau,[1] Microsoft PowerBI,[2] and Apache Spark[3] can be used to process and make sense of the data and present it

---

[1] https://www.tableau.com/

[2] https://powerbi.microsoft.com/

[3] https://spark.apache.org/

in a useful way to the human experts. Test automation tools like Junit,[4] GoogleTest,[5] and Eggplant[6] can be utilized to automate the experimental evaluation process, allowing to implement full test protocols and test cases in coding, and execute repetitive tests in an automated way, feeding their results to the data analytics tools for processing and visualization. In addition to scalar data, product testing can also produce unstructured data like texts and images. In this context, dedicated Natural Language Processing (NLP) and computer vision solutions can be used to make sense of the results. For instance, the authors in [18] train a dedicated object model to recognize structural defects in glass fiber-reinforced polymers (e.g., fiber breakage and fiber pull-out, cf. Fig. 4.3b). In [20], the authors utilize an NLP-based solution to process millions of text messaged captured by in-automobile communication network testing. They develop an anomaly detection algorithm to identify possible communication network anomalies, and then apply various visualization to highlight the communication nodes associated with those abnormalities, thus helping test engineers to achieve anomaly root cause reasoning. Recently, test engineers at idealworks have designed and implemented a battery of automated test protocols to evaluate the behavior of their iw.hub transport robots. The protocols are implemented to run on the robots' digital twins in the virtual environment, allowing to simulate different settings and constraints to measure and evaluate their performance (cf. Fig. 4.3a).

4.1.4   Training Phase

The training phase consists in providing the worker with the needed training skills to participate in the production process, and to provide the client with the needed skills to use the products once delivered. In this context, Virtual Reality (VR), Augmented Reality (AR), and digital twin technologies have been gaining increase popularity in multiple industrial sectors, providing realistic trainings which could improve employees' work performance and clients' experience. These so-called immersive technologies allow presenting complex and dangerous work scenarios in a virtual or augmented world, where people can train, learn new skills, and practice in a safe and informative environment. They can also help people retain what they learned, and provide them with an enhanced learning experience [21]. Leading organizations in the automotive, manufacturing, avionics, and logistics sectors are increasingly adopting immersive technologies to improve their learner experience [22]. BMW Group is a great example of immersive training, where employees are trained virtually in design and prototyping, setting-up Virtual Factory Layouts (VFLs) and performing their training scenarios in realistic digital twin

---

[4] https://junit.org/junit5/

[5] http://google.github.io/googletest/

[6] https://www.eggplantsoftware.com/

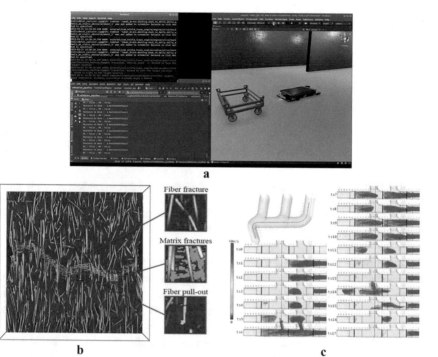

Fig. 4.3 Sample test data use cases (**b** and **c** are reported from [4]). (**a**). Sample run of an automated test for idealworks' iw.hub digital twin. (**b**). Identifying and recognize structural defects in glass fiber-reinforced polymers [18]. (**c**). Streamline flow visualization for the analysis of automotive exhaust, showing the gas flow direction within the exhaust system [19]

environments [23]. Dedicated training courses are offered to work on the assembly line, to train customer service employees, and to train safety managers [24]. On the other side of the Atlantic, engineers at Boeing and Ford Motor Company utilize physical props to enhance ergonomic evaluations and trainings (cf. Fig. 4.4a). Physical props are attached with smaller tracking markers to carefully align them with their digital twin counterparts in the virtual environment [25]. Human subjects train in the virtual world, by interacting with the physical props in the real world, providing them with an augmented sense of realism. The medical community has also made impressive strides in using immersive technologies as a training platform to expose beginner medical professionals to high-risk and difficult procedures [27]. For instance, the Children's Hospital Los Angeles use AI-powered VR simulations to train their medical students for emergency pediatric trauma situations [26] (cf. Fig. 4.4b). The hospital which spends around half a million dollars each year training doctors with practice models [28], is using immersive technologies to make its training program more efficient, while allowing medical students to practice more regularly in realistic conditions and learn based on existing cases re-created in the virtual world, while implementing unexpected scenarios to keep the students on their toes during the training [28].

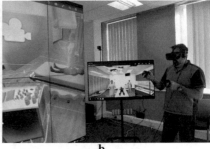

a   b

Fig. 4.4 Sample simulation and training systems for automotive (**a**) and medical (**b**) applications. (**a**). VR head mounted unit in front of a virtual environment for driver visibility simulation and training from Ford Motor Company [25]. (**b**). VR system for emergency pediatric trauma simulation and training at Children's Hospital Los Angeles [26]

Immersive technologies are specifically useful when performing high-risk training, in which learners put others at risk or are themselves at risk as they learn new tasks, such as pilots who learn to fly, or doctors who learn complex surgeries. Learning simulations and immersive training lowers those risks [21]. This is perfectly expressed by Allan Cook, managing director of Deloitte Consulting LLP: "If it's too difficult, too expensive or too dangerous to do the training in the real world, immersive training is a good fit". Immersive technologies also allow training for high-complexity scenarios that are difficult to recreate in the real world, such as pilots training for a plane crash, or doctors training for different kinds of medical emergencies in the operating room [21].

4.1.5   Service Phase

In the service phase of the production process, companies continue to track the product usage and users' experience even after the product leaves the production plant. Service data usually include information on flaws in product quality or defects in product design. Analyzing and mining such data is significant for companies to improve their after-sale service strategies and refine their product design and manufacturing, while maintaining good relationships with their clients [4]. This is specifically useful with high-tech consumer products like automobiles, mobile phones, and electronics which usually produce a large number of small faults that can only be discovered during usage [4]. A digital twin representation of the high-tech product would continuously monitor its behavior in the physical world, while predicting its remaining life, failure, and potential errors in the virtual world [29]. The concept of continuously monitoring a product during the service phased came initially from the airspace field, with the US Air Force Research

Laboratory developing digital twins to monitor the fatigue levels of US military aircraft [30]. Specific digital twin models were designed to monitor specific parts of the aircraft like detecting and monitoring the damages and faults in wing and chassis structure. The customer data, received in the form of reviews on the purchased products, can also be processed for sentiment analysis [31], topic extraction [32], and event mining [33], in order to better evaluate their satisfaction with certain products and take customer support actions accordingly (e.g., suggesting free support, or offering product replacements or refunds). For instance, the authors in [34] process automobile maintenance and repairing service reports to study the impact of different car usage habits on the service life of automobiles. This allows the automotive company to adjust and promote proper safety guidelines for its customers [4]. The authors in [35] introduce a visual approach to explore the temporal developments of car faults based on their maintenance data, in order to perform predictive maintenance and prevent critical failures.

4.2 Industrial Application Use Cases

4.2.1 Visibility

One of the mainstream use cases of using visual data in industry is evaluating the visibility of a person or a vision-enabled robot agent in a particular situation or given a particular posture. Cases in this category aim to answer the following questions [25]: "What can I see? What is blocking my visibility?" Evaluating the visibility of a stationary (non-moving) agent can be done using computer vision algorithms run on a stationary computer system. However, evaluating the visibility of a moving agent is more intricate, given the dynamicity of its surrounding environment [25]. Also, in the case of digital twin solutions, vision needs to be simultaneously evaluated in both the real world and in the virtual world, which adds another layer of complexity. For instance, many automotive manufacturers use VR (Virtual Reality) or AR (Augmented Reality) solutions to evaluate the visibility of drivers in newly designed autos (cf. Fig. 4.4a). In this context, the key to successful applications of visualization and visual analysis in the automotive industry is the industry's leading digitalization processes and data collection and support environments [4]. Digital renders and digital twins of the newest car models allow to evaluate the autos' properties, including the driver's front and rear visibility (cf. Fig. 4.5a). Engineers at the General Motors Design Lab study the influence of veiling glare from the instrument panels on the driver's windscreen and side window, calculating and rendering light reflections using dedicated algorithms [25]. This is especially important to better understand how the instrument panel affects driver visibility during night driving [36].

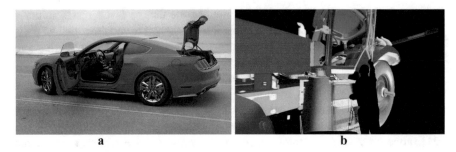

Fig. 4.5 Sample digital twin auto models. (**a**). High-resolution rendering of a Ford Mustang auto [25]. (**b**). A user checking the door handle location of a large tractor at Case New Holland [25]

4.2.2   Ergonomics

Another important use case is the person's interaction with the environment. Immersive technologies and digital twins are being increasingly used to evaluate the impact of physical tasks and the work environment on human workers. The question here is [25]: "How's someone going to posture themselves to do this task". Through its digital twin factory models, BMW Group's ergonomics engineers attempt to define the design criteria needed to optimize assembly line efficiency while minimizing worker effort. Using haptic feedback sensors and AR gloves, the engineers can estimate the forces needed to assemble certain parts of the car, and insert bolts and studs given different worker postures. The same immersive tools allow feedback to be obtained from the workers in the physical plant, which in turn feeds into the digital twin. At Case New Holland VR laboratories [37], ergonomics engineers use a power-wall display to evaluate the reachability of door handles in a vehicle buck. Their aim is to provide designs that allow drivers of many heights and strength to conformably reach the door handles and use them effectively (cf. Fig. 4.5b).

4.2.3   Packaging

Packaging concerns enfolding or placing tools or products in convenient locations or settings to prepare for their deployment, storage, transportation, sale, or usage [38]. In industrial settings, engineers often need to consider multiple packaging options to decide on the most convenient solution for the task at hand. Instead of physically implementing multiple packaging options which is labor intensive and costly, immersive technologies allow considering such options in the virtual world using VR, and superimposing them on the real world using AR, allowing the engineers and users to walk through the different possible scenarios to make sure they have a convenient solution. VR gives users a sense of space within the virtual world, while AR superimposes the virtual design on the real environment providing a sense of space within the real world. Whether it be a manufacturing plant floor, the interior of a car, a large room, or a small cockpit, packaging allows placing controls and tools at reasonable locations to best support the considered tasks [25]. For instance, BMW Group's uses its digital twin factory models to plan the organization of its large plant spaces, and help decide on the best placement of each component within the plant, from the assembly conveyor belt, to the supply storage pallets, trolleys, and everything in between (cf. Fig. 3.6). Similarly, engineers at PSA Peugeot Citroen use a three-sided cave VR environment to examine the possible placements of controls and instruments inside auto designs, to optimize interior car design and driver experience [39].

4.2.4   Realism

Developments in high-resolution graphics rendering and immersive technologies have improved so much that it is now possible to evaluate an object's look and feel interactively in a virtual environment. Improvements in lighting and material properties enable a near realistic product virtualization. Powerful rendering software include Unity,[7] Blender,[8] and NVIDIA Omniverse,[9] among others. NVIDIA's Omniverse is currently used by BMW Group and idealworks to develop photo-realistic tools, objects, machines, and full-fledged Virtual Factory Layouts (VFLs) as part of the SORDI dataset (cf. Figs. 4.6 and 4.7). Realism is of central importance when it comes

Fig. 4.6 Snapshot of the tools and supplies' packaging at the BMW factory digital twin at Regensburg

---

[7] https://unity.com/

[8] https://www.blender.org/

[9] https://www.nvidia.com/en-us/omniverse/

Fig. 4.7 Snapshot of BMW factory's digital twin at Regensburg, made of SORDI assets rendered using NVIDIA's Omniverse

to user experience, providing engineers, workers, and managers with a realistic look and feel of the virtual factory through its digital twin. It is essential to view geometry at real scale to understand the impact of problems or gaps of a digitized object, product or tool's form, function, and integration within its environment [40].

4.2.5 Storytelling

Many of the examples mentioned up until now have focused on the design of a particular object or product. Nonetheless, state of the art image rendering and digitalization technologies can also be used to tell stories in which an object or product is the central character. This is a core functionality of digital twins, where digitalized objects or products live their lives within the virtual world, in synchronization with their real counterparts in the physical world. The digital twin allows moving backward in time to view and study the object's state and behavior at a past timestamp. One can also fast-forward into the future, using trained deep learning and generative models to predict the object or product's state and behavior in the future [41, 42]. Design engineers can also preprogram specific scenarios to be executed through the virtual environment, in order to assess and evaluate an object or product's state and behavior within specific contexts [25].

References

1. 3 Pillar Global, *How Big Data is Transforming Industries in Big Ways.* 3 Pillar Global, 2022. https://www.3pillarglobal.com/insights/how-big-data-is-transforming-industries-in-big-ways/#:~:text=Better%20Decision%2DMaking,Improved%20Products%20and%20Services

2. Y. Wang et al., Visualization and visual analysis of multimedia data in manufacturing: A survey. Vis. Inform. **6**(4), 12–21 (2022)

3. Y. Zhao et al., Evaluating multi-dimensional visualizations for understanding fuzzy clusters. IEEE Trans. Vis. Comput. Graph. **25**(1), 1–10 (2019)

4. F. Zhou et al., A survey of visualization for smart manufacturing. J. Vis. **22**, 419–435 (2019)

5. S. Yin et al., A review on basic data-driven approaches for industrial process monitoring. IEEE Trans. Ind. Electron. **61**(11), 6418–6428 (2014)

6. F. Tao et al., Digital twin-driven product design framework. Int. J. Prod. Res. **57**(12), 3935–3953 (2019)

7. D. Coffey et al., Design by dragging: An interface for creative forward and inverse design with simulation ensembles. IEEE Trans. Vis. Comput. Graph. **19**(12), 2783–2791 (2013)

8. P. Ivson et al., Cascade: A novel 4D visualization system for virtual construction planning. IEEE Trans. Vis. Comput. Graph. **24**(1), 687–697 (2017)

9. A. Kratz et al., *Tensor Visualization Driven Mechanical Component Design.* IEEE Pacific Visualization Symposium (PacificVis'14), 2014. pp. 145–152

10. J. Weissenbock et al., *Fiberscout: An Interactive Tool for Exploring and Analyzing Fiber Reinforced Polymers.* IEEE Pacific Visualization Symposium (PacificVis'14), 2014. pp 153–160

11. H. Lipson, M. Kurman, *Fabricated: The New World of 3D Printing* (Wiley, New York, 2013), p. 291

12. P. Xu et al., Vidx: Visual diagnostics of assembly line performance in smart factories. IEEE Trans. Vis. Comput. Graph. **23**(1), 291–300 (2017)

13. T. Post et al., User-guided visual analysis of cyber-physical production systems. J. Comput. Inf. Sci. Eng. **17**(2), 021005 (2017)

14. F. Zhou et al., Visually enhanced situation awareness for complex manufacturing facility monitoring in smart factories. J. Vis. Lang. Comput. **44**, 58–69 (2017)

15. C. Arbesser et al., Visplause: Visual data quality assessment of many time series using plausibility checks. IEEE Trans. Vis. Comput. Graph. **23**(1), 641–650 (2017)

16. I. Somerville, *Software Engineering*, 10th edn. (Pearson, 2015), p. 816

17. K. Matkovic et al., Visual analytics for complex engineering systems: Hybrid visual steering of simulation ensembles. IEEE Trans. Vis. Comput. Graph. **20**(12), 1803–1812 (2014)

18. A. Amirkhanov et al., Visual analysis of defects in glass fiber reinforced polymers for 4DCT interrupted in situ tests. Comput. Graph. Forum **35**(3), 201–210 (2016)

19. P. Angelelli, H. Hauser, Straightening tubular flow for side-by-side visualization. IEEE Trans. Vis. Comput. Graph. **17**(12), 2063–2070 (2011)

20. M. Sedlmair et al., *Cardiogram: Visual Analytics for Automotive Engineers.* Proceedings of the SIGCHI Conference on Human Factors in Computing Systems, 2011. pp. 1727–1736

21. M. Pratt, *Top 5 Uses for VR in Learning and Development* (Tech Target, 2022). https://www.techtarget.com/searchhrsoftware/feature/Top-uses-for-VR-in-learning-and-development

22. Gartner, *Virtual Reality and Augmented Reality for Remote Workers* (Gartner Research, 2020). https://www.gartner.com/en/documents/3990221

23. Group, B, *A New Take on Vehicle Development* (BMW Press, 2022). https://www.bmw.com/en/events/nextgen/global-collaboration.html

24. Vrowl.io, *5 Examples of Virtual Reality Training in the Automotive Sector* (Vrowl.io, 2021). https://www.vrowl.io/5-examples-of-virtual-reality-training-in-the-automotive-sector/

25. L. Berg, J. Vance, Industry use of virtual reality in product design and manufacturing: A survey. Virtual Reality **21**, 1–17 (2017)

26. Games for Learning, *VR is Revolutionizing Trauma Training at Children's Hospital Los Angeles* (Games for Learning, 2023). https://www.g4li.org/virtual-reality/vr-is-revolutionizing-trauma-training-at-children-s-hospital-los-angeles.html
27. A. Liu et al., A survey of surgical simulation: Applications, technology, and education. Presence Teleop. Virt. **12**(6), 599 (2003)
28. Vrowl.io, *The 22 Best Examples of How Companies Use Virtual Reality for Training* (Vrowl. io, 2023). https://www.vrowl.io/the-22-best-examples-of-how-companies-use-virtual-reality-for-training/
29. Z. Zhu et al., *Visualization of the Digital Twin Data in Manufactoring by Using Augmented Reality.* 52th CIRP Conference on Manufactoring Systems, 2019. pp. 898–903
30. E. Tuege et al., Reengineering aircraft structural life prediction using a digital twin. Int. J. Aerosp. Eng. (2011). https://doi.org/10.1155/2011/154798
31. M. Fares et al., *Difficulties and Improvements to Graph-based Lexical Sentiment Analysis Using LISA.* IEEE International Conference on Cognitive Computing (ICC'19), 2019. pp. 28–35
32. S. Sarkissian, and J. Tekli, *Unsupervised Topical Organization of Documents Using Corpus-based Text Analysis.* International ACM Conference on Management of Emergent Digital EcoSystems (MEDES'21), 2021. pp. 87–94
33. M.A. Abebe et al., *Overview of Event-Based Collective Knowledge Management in Multimedia Digital Ecosystems.* International Conference of Signal Image Technology and Internet-based Systems (SITIS'17), 2017. pp. 40–49
34. S. Guo et al., EventThread: Visual summarization and stage analysis of event sequence data. IEEE Trans. Vis. Comput. Graph. **24**(1), 56–65 (2018)
35. Y. Chen et al., Sequence synopsis: Optimize visual summary of temporal event data. IEEE Trans. Vis. Comput. Graph. **24**(1), 45–55 (2018)
36. J. Hu et al., Study on the influence of opposing glare from vehicle high-beam headlights based on drivers' visual requirements. Int. J. Environ. Res. Public Health **19**(5), 2766 (2022)
37. Agriculture, N.H, *CNH Industrial Brand New Holland Collaborates with Microsoft and Touchcast at CES 2023 with a Metaverse Immersive Experience* (Lectura Press, 2023). https://lectura.press/en/article/cnh-industrial-brand-new-holland-collaborates-with-microsoft-and-touchcast-at-ces-2023-with-a-metaverse-immersive-experience/60304
38. A. Yoxall et al., Openability: Producing design limits for consumer packaging. Packag. Technol. Sci. **16**(4), 183–243 (2006)
39. P. Zimmermann, Virtual reality aided design. A survey of the use of VR in automotive industry, in *Product Engineering: Tools and Methods Based on Virtual Reality*, ed. by D. Talaba, A. Amditis, (Springer, Dordrecht, 2008), pp. 277–296
40. D. Bowman, R. McMahan, Virtual reality: How much imersion is enough? Comput. J. **40**(7), 36 (2007)
41. H. Noueihed et al., *Simulating Weather Events on a Real-world Map using Unity 3D.* Proceedings of the International Conference on Smart Cities and Green ICT Systems (SMARTGREENS'22), 2022. pp. 86–93
42. H. Noueihed et al., Knowledge-based virtual outdoor weather event simulator using Unity 3D. J. Supercomput. **78**(8), 10620–10655 (2022). https://doi.org/10.1007/s11227-021-04212-6

Chapter 5
Digital Images – The Bread and Butter of Computer Vision

The forth industrial revolution expresses a significant change in industry practices where state of the art digitalization technologies are bridging the gap between the physical and digital worlds, to allow improved productivity and services. In this context, manufacturing companies are increasingly using deep learning and computer vision solutions for product inspection, quality assurance, workplace safety, and factory automation through robotic vision [1]. For instance, a specific challenge that is of major importance to the BMW Group's Logistics department is the quality assurance of items delivered to the supply chain, and from the supply chain to the manufacturing plant. The department needs to make sure that (i) items acquired from external suppliers match the required specifications, and (ii) items that are delivered from the internal supply chain to the manufacturing plant remain in an acceptable state that matches the required specifications. In other words, the department is responsible for the quality of the items, starting from their acquisition from external suppliers, until their delivery to the manufacturing plant. To address this challenge, different processes are put in place, including manual check by human technicians, and automated checks by computer vision solutions. Another very important usage of computer vision in manufacturing is for automating quality inspection during the production process. Here, computer vision systems are trained to execute consistent quality checks according to industry standards to overcome variations across different human inspectors which might be subjective or inconsistent at times. Computer vision solutions also promote lean manufacturing, attempting to maximize productivity while minimizing waste by providing a visual data-driven approach to help in decision making.

***

At the core of computer vision lie digital images that allow the creation of visual data environments. Computer vision algorithms aim at identifying and recognizing salient objects or events in digital images, thus extracting useful information from digital images to perform automatic visual understanding.

J. Nassif et al., *Synthetic Data*, https://doi.org/10.1007/978-3-031-47560-3_5

In this context, how are digital images represented? What is the difference between feature-based and deep learning based computer vision solutions? Why is there a growing need for synthetic digital images in industry?

***.

We attempt to answer these questions in this chapter and in the remainder of this book…

5.1   Image Representation

Unlike classic image processing from a fixed database where each image is treated as an independent entity [2], image processing and computer vision on the supply chain and on the production pipeline in industry deals with integrated image items, including the images themselves as well as specification descriptions of the items. This is similar to images on the Web, where each image is contained within its host webpage which could underline a great deal of relevant information about the image itself [3]. Therefore, similarity to Web images, digital images in supply chain and manufacturing can be described not only by their visual features, but also by their textual specification descriptions and their related Web information in an integrated IoT (Internet of Things) setting, which are exploited to allow more effective processing.

In this context, we address digital image representation from different perspectives, ranging over low-level visual features, to textual features, joint word-image modeling, and 3D CAD processing.

5.1.1   Raster Images Versus Vector Images

Digital images can be organized in two main groups: (i) raster images, consisting of a set of pixels; and (ii) vector images made of geometric objects such as circles, triangles, rectangles, and polylines, etc. On the one hand, most existing approaches in the literature focus on the processing of raster images, e.g., [4, 5], which are produced by digital photo-taking cameras, and are capable of representing complex pictures having a variation of colors and shapes. On the other hand, vector images are becoming more popular in several application areas requiring the manipulation of small-size, resolution-independent, and simple images made of basic lines and shapes. These applications range over: industrial design applications (product and machine design using Computer Aided Design – CAD – solutions) [6], medical image annotation (adding basic shapes on top of medical images to identify organ tissues and tumors) [7, 8], geographic map annotation (highlighting special places and destinations on a map) [9, 10], and manipulating graph charts and simplifying accessibility to data and geometric shapes (producing simplified contour-based images to simplify data accessibility and navigation) [11, 12]. While the advantages and the practical applications of vector graphics highlight the importance of this category of images, e.g., [13, 14], yet, most existing image retrieval systems process vector images similarity to raster images [15], regardless of the properties offered by the former. This underlines two major challenges [16]: (i) undergoing expensive low-level feature selection and extraction, while disregarding the readily available geometric object features which can be extracted much more efficiently, and (ii) handling low-level features which are size and resolution dependent and which can affect retrieval quality, in contrast with vector graphics' features which are both resolution and size independent.

5.1.2   Low-Level Visual Features

Visual feature representation is the basis for content-based image processing and computer vision. A typical content-based system views each image as a collection of low-level visual features, and evaluates the relevance between images w.r.t.[1] their feature similarity [17]. Visual features can be grouped in three main categories: (i) color, (ii) texture, and (iii) shape. *Color descriptors* are used to represent the colors present in an image. Different color spaces exist in the literature such as CIE XYZ which attempts to generate a color model based on human eye color perception. Other color spaces include CIE RGB and CIELAB [2, 17]. Numerous color descriptors have also been proposed including color moments, color histogram, color coherence vector, color correlogram, etc. [18, 19]. The MPEG-7 multimedia metadata description standard has integrated more descriptors such as dominant color, scalable color, and color layout [20]. The use of color features usually depends on the nature of the images at hand. For example, for images which do not have an overall homogeneous color, the average or dominant color descriptors might not be very useful. On the other hand, domain knowledge such as color variance and color distribution over all images can be exploited to dynamically assign weights to image pixels [21], allowing to better compute color features. Color descriptors are most commonly used since they are comparatively simpler to process (compared with texture and shape features) and produce good enough results [2]. *Texture descriptors* are intended to capture the granularity and repetitive patterns of surfaces within in an image. They are usually made of spectral features, such as Gabor filtering [22] and wavelet transform [23], as well as statistical features such as the Wold features [24] and Tamura descriptors [25] (which are used in MPEG-7). Nevertheless, texture features are not as frequently used as their color counterparts, since they are more straightforwardly affected by image distortions and noise [17], and have been proven less effective on images where textures are not very structured and homogeneous (e.g., pictures of natural scenery) [26]. In [27], the authors propose a combined descriptor called Color Texture Moments (CTM), to integrate both color and texture characteristics in a compact form (using color moments and a Fourier

---

[1] With respect to.

transform based texture representation). Experimental results in [27] underline CTM's good performance w.r.t. its classic texture counterparts. *Shape descriptors* allow detecting different shapes and small salient objects in an image, and have been shown to be useful in many applications (especially when dealing with images of synthetic and man-made objects [17]). Shape descriptors include aspect ratio, circularity, consecutive boundary segments, Fourier descriptors, etc. [28]. MPEG-7 has adopted three main descriptors: 3-D shape descriptor derived from 3-D meshes of shape surface, region-based descriptor derived from Zernik moments, and the contour-based descriptor derived from the curvature scale space [20]. However, compared with color and texture, shape features are not so well defined and are not as commonly used [17], and remain relatively marginalized in many systems, e.g., [18, 21, 29].

5.1.3   High-Level Text-Based Features

While low-level visual features have been proven effective in content-based image processing [17], it is argued that the meaning (i.e., the semantics) of an image remains hardly self-evident [2]. Images which visual features are very similar to the query image may be very different from the query in terms of user interpretation and intended meaning. That is because human observers do not classically perceive an image in terms of pixel distributions, color patches, or surface features, but rather evaluate images at a higher semantic level, using lexical concepts (e.g., words or expressions) describing salient visual concepts in the image [30, 31]. Hence, a dedicated set of descriptors have been used to describe images on the Web, often referred to as: *high-level features* [2, 32], designating the textual content of the image. Textual descriptors include *tags*: which describe who and how many objects are found in a given picture, *location*: label (name or coordinates) of place where an image was taken, which can be utilized to allow geo-address comparison (using a geo-referenced ontology assigning geographic coordinates with place names [33]), *caption*: title of the image which is usually the most descriptive user-provided textual feature, providing a direct clue to the meaning and context of the image, *comments*: allowing a much greater variation of textual descriptions compared with the previous features, and they are especially useful when captions have not been provided by the user (publisher). The textual descriptors are then processed for feature representation, including word, phrase, sentence, and document level representations. These span over lexical form (origin of the term), semantic meaning (concept in a reference dictionary), part-of-speech tags (grammar category of the term), n-gram (word associations), syntactic structure (parse tree), and statistical features (e.g., contextual and co-occurrence term frequencies) [34, 35]. The features are subsequently represented as (one or multiple) high-level feature vector(s), where vector weights are computed using legacy term scoring methods developed in information retrieval.[2] While high-level features attempt to describe the semantics of the image (e.g., who, where, what, etc.) [17], however, their semantic descriptiveness hinges on on the quality of the nearby text portraying the meaning of the image.

Few recent methods have proposed expanding and enriching the textual descriptions of images, using techniques such as probabilistic image tagging (using the tagging logs to infer new tags, e.g., [36, 37]), and semi-supervised image labelling based on visual and Web contents (training various machine learning models to annotate new images based on a training image set with predefined labels, e.g., [38, 39]). While promising, yet the latter techniques require training data and training time, which are not always available.

5.1.4 Joint Word-Image Modeling

Word embeddings have lately proven to be a significant tool for the representation of word meanings in large text corpora. Their effectiveness relies on the distributional hypothesis that *words occurring in the same context carry similar semantic information.* As a result, few solutions have been developed to learn object and region embeddings from an image corpora. Word embeddings involve so-called *implicit semantics* (a.k.a. *latent semantics*) inferred from the statistical analysis of image tag names/labels in large corpora, following the basic idea that: documents (images) which have many labels in common are semantically closer than ones with fewer node labels in common [40]. *Implicit concepts* are *synthetic* concepts generated by extracting latent relationships between terms in a document or image collection, or by calculating probabilities of encountering terms, where the generated concepts do not necessarily align with any human-interpretable concept [41]. This is different from conventional concept-based semantic analysis, which utilizes *explicit concepts* representing *real-life* entities/notions defined following human perception (e.g., concepts defined within a conventional/non-conventional dictionary or knowledge base like WordNet or Wikipedia) [42]. For instance, the authors in [43, 44] study the co-occurrence of both visual and textual features using Latent Semantic Analysis (LSA) to produce a combined vectored data representation of both modalities. They extend the LSA to higher order to become applicable to more than two observable variables (visual and textual), and utilize cross-modal dependencies learned from corpora of tagged images to approximate the join distribution of the two variables. In [45], the authors adapt LSA and word2vec's skipgram and Contiguous Bag Of Words (CBOW) models to generate embeddings from object co-occurrences in images and subregions, and show that the produced embeddings improve typical object classification models by an average 3-to-4.5% top 1

---

[2]The standard *TF-IDF* (*Term Frequency – Inverse Document Frequency*) approach (or one of its variants) from the vector space model [183] is usually used, describing the number of times a term appears in a high-level feature (*TF*) compared with the number of times it appears in all entries of the feature (*IDF*).

accuracy. In [46], the authors introduce a Dual Path Recurrent Neural Network (DP-RNN) which processes images and sentences symmetrically by a deep learning model. Given an input image-text pair, the model reorders the image objects based on the positions of their most related words in the text. Similarity to extracting the hidden features from word embeddings, the model leverages the RNN to extract high-level object features from the reordered object inputs, producing similar representations in describing semantically related objects. The proposed approach produces state-of-the-art retrieval quality results compared with typical image retrieval techniques.

5.1.5   Multi-dimensional Image Feature Indexing

Multi-dimensional image indexing has also been investigated as a solution to combine and improve the representation of multiple image features, in order to allow faster access, processing, and retrieval. Most existing solutions fall into three main categories: (i) tree-based indexing, (ii) hashing-based indexing, and (iii) visual words based indexing. Tree-based indexing solutions sequentially partition the image search space and form hierarchical tree structures. The inner-nodes represent groups (clusters) of images or image regions, and the leaf nodes represent the images or the regions that are indexed. Image or region partitioning is conducted using dedicated image clustering algorithms, e.g., k-means and hierarchical k-means used for region-based cluster indexing in [47, 48]. Hash-based indexing solutions project image features from high dimensions to low dimensions using hash functions. Various approaches have been proposed, including Locality Sensitive Hashing (LSH) [49] build a family of spectral hashing functions where the probability of collision is higher for images that are close to each compared with those which are separate in the reduced dimensional space. Visual words based indexing solutions extract the local features from images, and quantize them into their closest visual words (codebook) based on a pre-learned training set [50, 51]. Then, a visual word-based vector is generated and is represented as an inverted file to allow for fast identification of images containing the visual word entries, and then fast feature processing and similarity computation.

5.1.6   3D CAD Processing

Digitizing a real-world environment, like a BMW factory plant, requires 3D digital imaging, so-called 3D Computer Aided Design (CAD) processing. This means that the first step in creating a digital twin factory is to produce an accurate 3D scene with 3D models that can be used to run simulations. The process starts by doing a 3D scan of the factory using advanced Lidar sensors, producing a point cloud representation of the different objects in the scene. Although the resulting point cloud

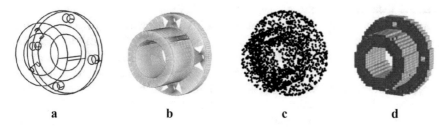

a   **b**   **c**   **d**

Fig. 5.1 3D CAD image processing [52]. (**a**) 3D CAD. (**b**) Mesh. (**c**) Point Cloud. (**d**) Voxel

is a 3D representation of the plant, yet it cannot be used for simulations because it is simply a collection of points in the 3D space. In order to do any kind of useful simulation, the 3D representation must introduce the concepts of objects and relationships between different objects, which cannot be achieved with a point cloud representation. This requires transforming the point cloud representation into a 3D scene made of 3D models, which is achieved in two stages [52]: (i) mesh file conversion, and (ii) voxel and point cloud transformation (cf. Fig. 5.1). First, the collected 3D CAD is transformed into a mesh file. Mesh files denote the shapes of 3D CAD models in multiple triangles and store triangular information as face and node information. Tools like FreeCAD (2020) [53] can be used to automate the mesh file conversion process. After converting to a mesh file, a 3D grid is constructed using the maximum and minimum values of the triangle's node coordinates (x, y, z) to convert mesh-to-voxel. They are divided into voxel grid sizes. In a split 3D grid, a voxel is formed in that grid when the triangle of the mesh file intersects in the grid [54]. Third, the number of output points is produced to convert mesh-to-point cloud. Then, a weighted random sampling method is used to select triangles in the 3D mesh by n points. Within each randomly selected triangle, one point in a random coordinate is generated using the triangle's center of gravity method. Mesh is lastly transformed to point cloud after repeating this point-generation process multiple times [55].

5.2   Need for Synthetic Images

5.2.1   Shift in Computer Vision Paradigms

Computer vision tasks are traditionally achieved following a 2-steps process: (i) applying a hand-crafted feature extraction algorithm [56] according to the features described in the Sect. 5.1, followed by (ii) training a traditional Machine Learning (ML) model on the extracted features [57]. In the last 10 years, Deep Learning (DL) techniques have increasingly surpassed their traditional 2-step process counterparts in terms of both accuracy and inference time [58, 59]. Although these DL models learn to extract relevant features from input images in an end-to-end manner, training them needs capturing, storing, and labelling large amounts of images in comparison to traditional ML approaches [60]. More specifically, the acquisition of large image datasets, mainly in industrial settings (e.g., factories) is becoming increasingly challenging and critical due to the following reasons [61]. First, excessive human effort is needed for manual image capture, image preprocessing (e.g. cropping, filtering out noisy images) and image annotation (e.g. bounding boxes and pixel-wise segmentation). Second, these tasks and more specifically image annotation, are extremely prone to errors and are highly subjective. For example, target objects can be incorrectly annotated according to the human annotator's knowledge and experiences, especially when it comes to overlapped and close objects or to similar but different scale assets [62]. Third, capturing images inside industrial locations and factories plants can be difficult due to limited access for security (e.g. innovation and high security areas), privacy [63] (e.g. human workers along the assembly lines), or safety reasons (e.g. painting or assembly lines, etc.).

5.2.2 Need for Synthetic Images

Synthetic image generation can tackle the above challenges while acquiring a large image dataset with the required properties, e.g., multi-modal captures for various scene conditions, lighting, resolution, and object obfuscation ratio, among others [64]. Image modality refers to the method in which an image is captured or generated, where each modality represents a different type of information, e.g. bounding box image, segmentation, depth, etc. [61]. Using a graphics rendering tool such as NVIDIA's Omniverse, Blender, Unreal, or Unity, it is possible to control every aspect that affects the image, ranging over lighting, camera position, assets distribution and animation, and noise, among others. Also, renderers automatically generate numerous image and object annotations based on the selected modalities [65]. Nonetheless, synthetic image generation highlights one main challenge: the "reality gap", i.e., the difference between real and synthetic images [66, 67]. The objective of synthetic image generation is to reduce the reality gap, such that DL models trained on the synthetic data can be straightforwardly applied on real images while maintaining high accuracy results, without the need to train them on actual real images which are oftentimes unavailable or insufficient.

5.3   Computer Vision Datasets

Multiple computer vision datasets have been developed recently and have been used in many fields. We categorize them under (i) general purpose image datasets, (ii) industrial image datasets, and (iii) general purpose synthetic image datasets, and (iv) industrial synthetic image datasets.

5.3.1   General Purpose Image Datasets

CIFAR-100 [68] and MS COCO [69] are well known general purpose computer vision datasets. The CIFAR-10 dataset is made of 60 k 32 × 32 color images organized in 10 classes, with 6000 images per class. It is divided into 50 k of training images and 10 k of test images. MS COCO consists of more than 200,000 images covering more that 90 asset classes of daily life facilities, various fauna and flora, as well as transportation means. ImageNet [70] is another common image database described according to the WordNet hierarchy with over 100,000 synsets, and an average of 1000 images per synset. A synset in Wordnet stands for a semantic concept describing a set of synonymous terms and their definition. The concepts are connected together through multiple types of semantic relationships like hypernym, hyponym, meronym, among others [71]. For instance, synset *car* includes synonymous terms *car, auto, automobile*, is connected to concept *vehicle* through an inheritance/isA relationship (*car*-isA-*vehicle*), and is connected to concept engine through a meronym/partOf relationship (*car*-hasA-*engine*). Altogether, ImageNet provides tens of millions of well sorted images organized according to the WordNet semantic network. Nonetheless, ImageNet does not seem to be easily scalable anymore since it was human-annotated over multiple years by different groups of people, resulting in many inconsistencies in labelling, where different class labels for the same object class [72]. Also, ImageNet does not comply with the recent data privacy and security policies [73], namely not revealing human faces. This is very important in commercial and industrial applications where people's identities need to be kept private during image processing and object recognition. To handle this problem, the creators of ImageNet are reproducing the dataset by running multiple object detection and image processing scripts allowing to obscure sensitive information like people's faces, geographic sign texts, store names, etc. This is not an easy task and requires manual effort and supervision to maintain the overall usability and effectiveness of the previous dataset version [74].

5.3.2   Industrial Image Datasets

Many industrial image datasets have been developed and utilized for specific industrial applications. For instance, the authors in [75] describe an industrial image dataset dedicated for casting manufacturing products. Casting is a manufacturing process in which a warm liquid material is poured into a mold, containing a hollow cavity of the desired shape, where the liquid metal is allowed to cool down to solidify. The dataset includes 7348 images of 300 × 300 pixels top views of the submersible pump impeller (cf. samples in Fig. 5.2a). Two classes are considered: (i) defective object and (ii) normal object, making this dataset well suited for building binary classifiers [75]. Another industrial dataset is GC10-DET [76], sometimes referred to as the Defect dataset. It consists of 3570 gray-scale images describing

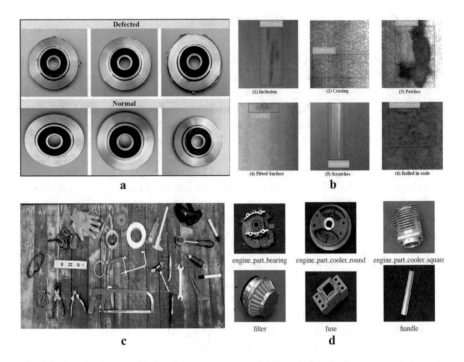

Fig. 5.2 Samples from real industrial image datasets. (**a**) Sample images from the pump impeller Casting dataset [75]. (**b**) Sample images from the GC10-DET metal surfaces dataset [76]. (**c**) Sample images from ITD industrial tools dataset [80]. (**d**) Sample images from MVTec_ITODD industrial objects dataset [82]

ten different types of metal surface defects (e.g., *punching*, *weld line*, *crescent gap*, *water spot*, *oil spot*, among others, cf. samples in Fig. 5.2b). Distinguishing between such types of defects is of central important in the steel production industry, and can significantly contribute to preventing malfunction, as well as identifying defects on a variety of essential materials found in factories and processing plants. The authors in [77] introduce the MCuePush dataset consisting of 1243 real images describing magnetic tile defects. The images are labelled according to 6 defect categories (i.e., *blowhole*, *crack*, *fray*, *break*, *uneven*, and *defect_free*). In [78], the authors describe SDNET2018, a dataset of 56,000 images labeled as *cracked* and *non-cracked* concrete bridge decks, walls, and pavements, where cracks are as narrow as 0.06 mm and as wide as 25 mm. The dataset includes artifacts like shadows, surface roughness, scaling, edges, holes, and background debris, making the recognition task more challenging. Another recent dataset for solar cell defect detection is described in [79], consisting of 2624 images of *functional* and *defective* solar cell surfaces where the defective class covers a variety of defect types (e.g., material defect finger interruptions, microcrack, etc.). While the above mentioned datasets are interesting in their own respect and application use case, nonetheless, we remark that they consider very specific and narrow industrial applications.

Differently from the above, the authors in [80] describe a more comprehensive Industrial Tool Dataset (ITD) to detect different shaped tools that are found in different manufacturing industries. The dataset consists of 11,000 images manually labelled by mechanical engineers according to 8 categories (*adhesive tools, fastener tools, protection tools, cutting tools*, etc., cf. Fig. 5.2c). The dataset focuses on small-size assets found on a tool shell or on a workshop table. In a comparable study, the authors in [81] describe the Logistics Objects in Context dataset (LOCO), describing logistics specific objects such as *pallets, pallet trucks, forklifts* and *small pallet loaders*. The dataset consists of 5593 images manually annotated, where people's faces are pixelated when they occur to preserve their identities. According to the authors in [81], using different cameras and recording methods resulted in many blurry images that require additional data cleaning and image post-processing before performing object recognition.

Differently from the above mentioned datasets which are made of typical 2D raster images, the MVTec Industrial 3D Object Detection Dataset (MVTec ITODD) [82] is designed for object detection and pose estimation in 2D and 3D space. It consists of 3500 labeled images of industrial objects categorized under 28 object classes (e.g., *cap, filter, fuse, screw, handle*, etc., cf. Fig. 5.2d). The authors use a multi-shot, wide-baseline 3D stereo sensor, providing a range (Z) image, X and Y images, as well as a grayscale image with the same viewpoint as the range image. The sensor uses multiple random projected patterns and reconstructs the scene using a space-time stereo approach with an accuracy of around 100 μm. The authors use three high-resolution cameras ($\approx$ 8 MP, f = 50 mm) to capture the images.

5.3.3 General Purpose Synthetic Image Datasets

Various synthetic datasets have been recently developed in order to augment real data when training deep learning computer vision models. For instance, the SYNTHIA dataset [83] consists of 213,400 synthetic images (1280 × 980 pixels) of urban images generated based on random snapshots and video frame sequences. The images are automatically labelled according to 13 classes (e.g., *building, road, sidewalk, fence, pedestrian*, etc.) in urban driving scenarios with changing seasons, weather conditions, and lighting conditions (cf. Fig. 5.3a). Annotation is conducted using pixel-level segmentation, using a dedicated Fully Convolutional Network (CNN) model [84] to perform the segmentation task. The authors train various deep learning models using (i) real data only, (ii) SYNTHIA synthetic data only, (iii) both real and SYNTHIA synthetic data. While the accuracy of the models trained on real data surpasses those trained on synthetic data, nonetheless, the difference in accuracy levels is not drastic and ranges between 5%-to-8% on average. According to the authors in [83], this shows that training on SYNTHIA produces good enough results to perform urban object recognition. In addition, and most importantly, models trained on both real and synthetic data combined, produce maximum accuracy levels, with an average increase of 0.5%–5% compared with training on real data

Fig. 5.3 Samples from general-purpose synthetic image datasets. (**a**) Samples images from the SYNTHIA urban driving dataset [83]. (**b**) Sample images from the FAT household objects dataset [87]

only. Another general purpose synthetic dataset is SIDOD [85], consisting of 144,000 pairs of stereo images, using 18 camera views from 3 virtual scenarios with randomly selected household items from the Yale-CMU-Berkeley (YCB) daily life dataset [86]. SIDOD images are generated using NVIDIA's Deep Learning Data Synthesizer (NDDS) [87], which is built on top of the Unreal Engine. It renders images with a high frame rate, including multiple features such as: depth, stereo, full rotation, 3D pose, occlusion, segmentation, and flying distractors. It is intended for object detection, pose estimation, and tracking applications. SIDOD's main scene assets and distractors are randomized at each frame instead of capturing a random YCB asset during its falling animation in a static virtual environment or background. However, SIDOD includes assets found in households such as *bowl* and *tomato soup can*, making it unsuitable for industrial applications. Another similar dataset is FAT [87] consisting of 60,000 synthetic images of household objects including 21 categories from the YCB dataset (e.g., *banana*, *powder drill*, *mug*, *bowl*, etc.) [86]. For each image, the authors provide the 3D poses, per-pixel class segmentation, and 2D/3D bounding box coordinates for their contained object models. The object models are combined with backgrounds of complex composition and high graphical quality to produce photorealistic images (cf. Fig. 5.3b). While the above datasets provide high quality synthetic images, nonetheless, they target general purpose objects and scenarios (i.e., urban driving with SYNTHIA, and household objects with SIDOD and FAT), making them unsuitable for industrial applications.

5.3.4   Industrial Synthetic Image Datasets

Few synthetic datasets exist to date for generating industrial images. The most notable is T-LESS [88] consisting of 10,000 3D synthetic images categorized under 30 industry-relevant objects (consisting of different kinds of *light switches*, *light bulb connectors*, and *electric connectors*, cf. Fig. 5.4). Each object is represented by two 3D models: the first one is created using CAD (Computer Aided Design), and the second one is semi-automatically reconstructed from RGB-D (Depth) images using fast fusion [89], a volumetric 3D reconstruction system. The objects show symmetries and mutual similarities in shape or size, where some of the objects are parts of others. T-LESS synthetic images vary from simple to complex scenes with multiple instances and a high amount of occlusions and clutter. T-LESS image scenes are exported as one single mesh without annotations, in spite of the benefits 3D scanning in providing highly realistic meshes. Hence, each object must be scanned independently and merged within the scene to re-construct the synthetic image, which is a significantly time-consuming process [61]. The authors in [88] report their dataset's performance in performing 6D pose estimation and deduce that there is significant room for improvement, especially in cases with significant object occlusion.

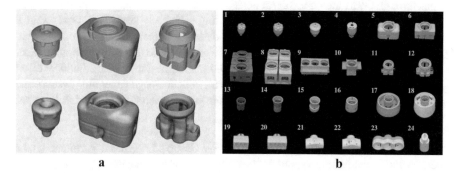

Fig. 5.4 Samples from the T-LESS industrial synthetic image dataset [88]. (**a**) Sample 3D CAD object models from T-LESS. (**b**) Sample images from the T-LESS dataset

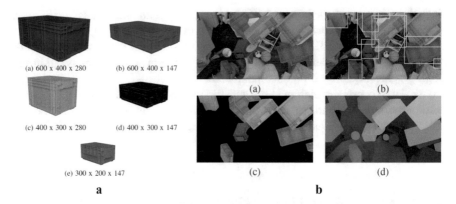

Fig. 5.5 Samples rendered images of KLT boxes generated according to the approach in [81]. (**a**) Sample 3D object models of KLT boxes. (**b**) For each rendered image (a), the system automatically generates bounding boxes (b), panoptic segmentation masks (c), as well as depth maps (d)

To ensure the scalability of their synthetic dataset, the authors in [81] put forward an automated pipeline for synthetic image generation of industrial objects using Blender. The pipeline consists of three main phases. First, a background is composed of many random 3D objects as well as a background image to fill the void gap between the objects. Second, target objects and object-alike distractors are randomly placed within the camera view. Third, lighting and camera are added randomly using a Mixed-Lighting Illumination (MLI) approach: combining global and local light sources to automatically create a diverse illumination of the scene. The authors randomize the number, type, location, rotation and material of target objects. In addition, the authors make use of Object Relation Modelling (ORM) to apply predefined relations between target objects when placing them in 3D space. To do so, they manually create relation files by recording the relative translation and rotation of one target object to another before the image is generated. This spatial relation can later be randomly applied during foreground generation. If an intersection is detected when applying a relation, the related object is deleted. Experimental results in [81] show that considering object relations through ORM increases object detection accuracy when applied on real images. The authors evaluate their pipeline by detecting differently sized KLT boxes in real images, and show that that real image-based detectors outperform their proposed synthetic image based detectors (cf. Fig. 5.5). The authors state that their generation pipeline, in its current form, highly depends on full randomization, where additional effort is needed to analyze Blender's capabilities as a synthetic image generation system, and improving their pipeline to minimize the domain gap between synthetic and real images.

***

Going back to the challenges faced by BMW Group's Logistics department: (i) the need to verify that items acquired from external suppliers match the required specifications, and (ii) the need to verify that items delivered from the internal supply chain to the manufacturing plant remain in an acceptable state and match the required specifications; none of the mentioned datasets can be straightforwardly used to address the aforementioned challenges. This highlights the need for a full-ledged industrial synthetic objects dataset that is easily extensible and scalable, using a state of the art graphics rendering environment that can generate photorealist images where objects share realistic properties that parallel their real-world counterparts. This is where BMW Group's SORDI (Synthetic Object Recognition Dataset for Industries) comes into play, designed for smart robot object detection and recognition in industries and manufacturing plants. SORDI uses the powerful NVIDIA Omniverse engine to create realistic visual patterns and illumination, while simulating the objects' physical properties such as weight, surface texture, extreme lighting, and drag. We describe SORDI's creation pipeline, assets, and their properties in the following chapter of this book.

References

1. G. Boesch, *Computer Vision in Manufacturing – The Most Popular Applications in 2023*. Viso. ai, 2023. https://viso.ai/applications/computer-vision-in-manufacturing/
2. R. Datta, D. Joshi, J. Li, J.Z. Wang, Image retrieval: Ideas, influences and trends of the new age. ACM Computer Surveys **40**(2), 1–60 (2008)
3. X. He et al., *ImageSeer: Clustering and Searching WWW Images Using Link and Page Layout Analysis*. Microsoft Technical Report – MSR-TR-2004-38, 2004
4. S. Wagenpfeil et al., Fast and effective retrieval for large multimedia collections. Big Data and Cognitive Computing **5**(3), 33 (2021)
5. J. Jagtap, N. Bhosle, A comprehensive survey on the reduction of the semantic gap in content-based image retrieval. Intl. J. Appl. Pattern Recogn. **6**(3), 254–271 (2021)
6. D. Madsen, D. Madsen, *Engineering Drawing and Design*, Cengage Learning, 6th edn (2016), 1680 p
7. E. Kim, et al., *A Hierarchical SVG Image Abstraction Layer for Medical Imaging*. Society of Photo-Optical Instrumentation Engineers (SPIE) Conference, 2010. 7628, 7
8. K. Salameh, et al., *SVG-to-RDF Image Semantization*. 7th International SISAP Conference, 2014. pp. 214–228
9. D. Li, et al., Shape similarity computation for SVG. Int. J. Comput. Sci. Eng. **6**(1/2) (2011)
10. Z.R. Peng, C. Zhang, The roles of geography markup language (GML), scalable vector graphics (SVG), and Web feature service (WFS) specifications in the development of Internet geographic information systems (GIS). J. Geogr. Syst. **6**, 95–116 (2004)
11. J. Tekli et al., Evaluating touch-screen vibration modality toward simple graphics accessibility for blind users. Intl. J. Human Comp. Stud. (IJHCS) **110**, 115–133 (2018)
12. C. Engel et al., SVGPlott: An accessible tool to generate highly adaptable, accessible audio-tactile charts for and from blind and visually impaired people. PETRA **2019**, 186–195 (2019)
13. H. Gaudenz et al., VIAN: A visual annotation tool for film analysis. Computer Graphics Forum **38**(3), 119–129 (2019)
14. Spindler M., et al., *Translating Floor Plans into Directions*. Proceedings of the 13th international conference on computers helping people with special needs, 2012. Linz, Austria
15. K. Jiang, et al., *Information Retrieval through SVG-based Vector Images Using an Original Method*, in Proceedings of IEEE International Conference on e-Business Engineering (ICEBE'07) 2007. pp. 183–188
16. K. Salameh et al., Unsupervised knowledge representation of panoramic dental X-ray images using SVG image-and-object clustering. Multimedia Syst. (2023). https://doi.org/10.1007/s00530-023-01099-6
17. Y. Liu, D. Zhang, G. Lu, W.-Y. Ma, A survey of content-based image retrieval with high-level semantics. Pattern Recogn. **40**(1), 262–282 (2006)

18. P.L. Stanchev, D. Green Jr., B. Dimitrov, High level color similarity retrieval. Intl. J. Inform. Theory Appl. **10**(3), 363–369 (2003)
19. Y. Liu, D. Zhang, G. Lu, W.-Y. Ma, Region-based image retrieval with perceptual colors. Proc. Pacific-Rim Multi Conf (PCM), 931–938 (2004)
20. B.S. Manjunath, *Introduction to MPEG-7* (Wiley, New York, 2002), p. 412
21. K.A. Hua, K. Vu, J.H. Oh, *Proceedings of the 7th ACM International Multimedia Conference (ACM MM'99).* A Flexible and Efficient Sampling-based Image Retrieval Technique for LArge Image Databases, SamMatch, pp. 225–234
22. B.S. Manjunath, W.Y. Ma, Texture features for browsing and retrieval of image data. IEEE Trans. Pattern Anal. Mach. Intell. **18**(8), 837–842 (1996)
23. J.Z. Wang, J. Li, G. Wiederhold, SIMPLIcity: Semantics-sensitive integrated matching for picture libraries. IEEE Trans. Pattern Anal. Mach. Intell. **23**(9), 947–963 (2001)
24. F. Liu, R.W. Picard, Periodicity, directionality, and randomness: World features for image modelling and retrieval. IEEE Trans. Pattern Anal. Mach. Intell. **18**(7), 722–733 (1996)
25. H. Tamura, S. Mori, T. Yamawaki, Texture features corresponding to visual perception. IEEE Trans. Syst. Man Cybern. **8**(6), 460–473 (1978)
26. W.K. Leow, S.Y. Lai, *Scale and Orientationi-invariant texture matching for image retrieval* in Pietikainen (Ed.) texture analysis in machine vision, 2000. Pp. 151-163, world scientific. Dermatol. Sin
27. H. Yu et al., *Color Texture Moments for Content-based Image Retrieval,* in Proceedings of the International Conference on Image Processing (ICIP), 2002. pp. 24–28
28. R. Mehrotra, J.E. Gary, Similar-shape retrieval in shape data management. IEEE Comp. **28**(9), 57–62 (1995)
29. V. Mezaris et al., *An Ontology Approach to Object-based Image Retrieval.* International Conference on Image Processing (ICIP'03), vol. 2, 2003. pp. 511–514,
30. J.A. Black Jr., K. Kahol, P. Kuchi, G. Fahmy, S. Panchanathan, *Characterizing the High-Level Content of Natural Images Using Lexical Basis Functions* (SPIE, Human Vision and Electronic Imaging VIII, 2003), pp. 378–391
31. Y. Chen et al., *Content-based Image Retrieval by Clustering.* Proceedings of the ACM International Conference on Multimedia Information Retrieval (MIR'03), 2003. pp. 193–200
32. X. Li, et al., Socializing the semantic gap: A comparative survey on image tag assignment, refinement, and retrieval. ACM Comput. Surveys. **49**(1): 14:1–14:39 (2016)
33. J. Tekli et al., *Toward Approximate GML Retrieval Based on Structural and Semantic Characteristics,* in Proceedings of the International Conference on Web Engineering (ICWE'09), 2009. pp. 16–34
34. M. Fares et al., Unsupervised word-level affect analysis and propagation in a lexical knowledge graph. Elsevier Knowl. Based Syst. **165**, 432–459 (2019)
35. V. Soares et al., Combining semantic and term frequency similarities for text clustering. Knowl. Inf. Syst. **61**(3), 1485–1516 (2019)
36. V. Papapanagiotou et al., Improving concept-based image retrieval with training weights computed from tags. ACM Trans. Multim. Comput. Commun. Appl. **12**(2), 32:1–32:22 (2016)
37. M. Ruocco, H. Ramampiaro, Event-related image retrieval: Exploring geographical and temporal distribution of user tags. Intl. J. Multim. Inform. Retr. **2**(4), 273–288 (2013)
38. L. Ma et al., Learning efficient binary codes from high-level feature representations for multi-label image retrieval. IEEE Trans. Multimed. **19**(11), 2545–2560 (2017)
39. B. Madduma, S. Ramanna, Image retrieval based on high level concept detection and semantic labelling. Intellig. Dec. Technol. **6**(3), 187–196 (2012)
40. J. Tekli, An overview on XML semantic disambiguation from unstructured text to semi-structured data: Background, applications, and ongoing challenges. IEEE Trans. Knowl. Data Eng. (IEEE TKDE) **28**(6), 1383–1407 (2016)
41. X. Yi, J. Allan, *A comparative study of utilizing topic models for information retrieval,* in Proceedings of the 31st European Conference on IR Research (ECIR'09), 2009. pp. 29–41
42. E.M. Voorhees, *Using Wordnet to Disambiguate Word Senses for Text* Retrieval, in Proceedings of the 16th Annual International ACM SIGIR Conference on Research and Development in Information Retrieval, 1993. pp. 171–180

43. S. Nikolopoulos et al., High order pLSA for indexing tagged images. Signal Process. **93**(8), 2212–2228 (2013)
44. S. Giouvanakis, C. Kotropoulos, *Saliency Map Driven Image Retrieval Combining the Bag-of-Words Model and PLSA*. International Conference on Digital Signal Processing (DSP'14), 2014. pp. 280–285
45. M. Treder, et al., *Deriving visual semantics from spatial context: An adaptation of LSA and Word2Vec to generate object and scene embeddings from images*. CoRR abs/2009.09384, 2020
46. T. Chen, J. Luo, *Expressing objects just like words: Recurrent visual embedding for image-text matching*. AAAI Conference on Artificial Intelligence (AAAI'20), 2020. pp. 10583–10590
47. S. Hussain, M. Haris, A K-means based co-clustering (kCC) algorithm for sparse, high-dimensional data. Expert Syst. Appl. **118**, 20–34 (2019)
48. X. Wang et al., High-dimensional data clustering using K-means subspace feature selection. J. Netw. Intell. **4**(3), 80–87 (2019)
49. O. Durmaz, H.S. Bilge, Fast image similarity search by distributed locality sensitive hashing. Pattern Recogn. Lett. **128**, 361–369 (2019)
50. H. Sun, et al., *Commodity Image Classification Based on Improved Bag-of-Visual-Words Model*. Complexity, 2021. 2021: 5556899:1–5556899:10
51. M. Saini, S. Susan, Bag-of-visual-words codebook generation using deep features for effective classification of imbalanced multi-class image datasets. Multim. Tools Appl. (MTAP) **80**(14), 20821–20847 (2021)
52. S. Yoo, N. Kang, Explainable artificial intelligence for manufacturing cost estimation and machining feature visualization. Expert Syst. Appl. **183**, 115430 (2021)
53. FreeCAD, Accessed April 2023. https://www.freecadweb.org/
54. A. Adam, *Mesh Voxelisation*. MathWorks, 2013. https://www.mathworks.com/matlabcentral/fileexchange/27390-mesh-voxelisation
55. D. Iglesia, *3D Point Cloud Generation from 3D Triangular Mesh*. 2017. https://medium.com/@daviddelaiglesiacastro/3d-point-cloud-generation-from-3dtriangular-mesh-bbb602ecf238
56. J. Tekli, An overview of cluster-based image search result organization: Background, techniques, and ongoing challenges. Knowl. Inf. Syst. **64**(3), 589–642 (2022)
57. S. Paisitkriangkrai et al., Effective semantic pixel labelling with convolutional networks and conditional random fields. IEEE Conf. Comput. Vision Pattern Recogn. Workshops, 36–43 (2015)
58. S. Mittal, S. Vaishay, A survey of techniques for optimizing deep learning on gpus. J. Syst. Archit. **99**, 101635 (2019)
59. R. Al Sobbahi, J. Tekli, Comparing deep learning models for low-light natural scene image enhancement and their impact on object detection and classification: Overview, empirical evaluation, and challenges. Signal Process. Image Commun. **109**, 116848 (2022)
60. C. Sun, et al., *Revisiting Unreasonable Effectiveness of Data in Deep Learning Era*, in Proceedings of the IEEE International Conference on Computer Vision, 2017. pp. 843–852
61. C. Abou Akar, et al., *Synthetic Object Recognition Dataset for Industries*. International Conference on Graphics, Patterns and Images (SIBGRAPI'22), 2022. pp. 150–155
62. M. Ayle, J. Tekli, et al., *Bar—A reinforcement learning agent for bounding-box automated refinement*. Proceed. AAAI Conf. Artif. Intell. **34**(03), 2561–2568 (2020)
63. J. Tekli, et al., *A framework for evaluating image obfuscation under deep learning-assisted privacy attacks*. 17th International Conference on Privacy, Security and Trust (PST'19), 2019. pp. 1–10
64. E. Jurado et al., Towards the generation of synthetic images of palm vein patterns: A review. Informat. Fusion **89**, 66–90 (2023)
65. Unity, *Materials, Shaders, and Textures.*. https://docs.unity3d.com/560/Documentation/Manual/Shaders.html, Accessed April 2023
66. J. Tobin, et al., *Domain Randomization for Transferring Deep Neural Networks from Simulation to the Real World*. IEEE/RSJ international conference on intelligent robots and systems (IROS'17), 2017. pp. 23–30

67. W. Chen, et al., *Contrastive syn-to-real generalization.* arXivpreprint arXiv:2104.02290, 2021
68. A. Krizhevsky, et al., *The cifar-10 and cifar-100 dataset.* https://www.cs.toronto.edu/~kriz/cifar.html, 2021
69. T. Lin, et al., *Microsoft Coco: Common Objects in Context.* European conference on computer vision. Springer, 2014. pp. 740–755
70. J. Deng, et al., *Imagenet: A large-scale hierarchical image database.* IEEE Conference on Computer Vision and Pattern Recognition (CVPR'09), 2009. pp. 248–255
71. J. Tekli et al., Full-fledged semantic indexing and querying model designed for seamless integration in legacy RDBMS. Data Knowl. Eng. **117**, 133–173 (2018)
72. L. Beyer, et al., *Are we done with imagenet?* arXiv preprint arXiv:2006.07159, 2020
73. J. Whitaker, *The fall of imageNet.* https://towardsdatascience.com/the-fall-of-imagenet-5792061e5b8a, 2021
74. K. Johnson, *ImageNet Creators Find Blurring Faces for Privacy Has a 'Minimal Impact on Accuracy.* 2022. https://venturebeat.com/2021/03/16/imagenet-creators-find-blurring-facesfor-privacy-has-a-minimal-impact-on-accuracy
75. I. Apostolopoulos, M. Tzani, *Industrial Object, Machine Part and Defect Recognition Towards Fully Automated Industrial Monitoring Employing Deep Learning the Case of Multilevel vgg19.* arXiv preprintarXiv:2011.11305, 2020
76. X. Lv et al., Deep metallic surface defect detection: The new benchmark and detection network. Sensors **20**, 1562 (2020)
77. Y. Huang, et al., *Surface Defect Saliency of Magnetic Tile.* IEEE 14th International Conference on Automation Science and Engineering (CASE'18), 2018. pp. 612–617
78. M. Maguire, et al., *SDNET2018: A Concrete Crack Image Dataset for Machine Learning Applications.* Utah State University Libraries, 2018. https://doi.org/10.15142/T3TD19
79. S. Deitsch et al., Segmentation of photovoltaic module cells in uncalibrated electroluminescence images. Machine Vision Appl. (Springer) **32**(4), 84 (2021)
80. C. Luo et al., A benchmark image dataset for industrial tools. Pattern Recogn. Lett. **125**, 341–348 (2019)
81. C. Mayershofer, et al., *Towards Fully-Synthetic Training for Industrial Applications.* International Conference on Logistics, Informatics and Service Sciences (LISS'20), 2021. pp. 765–782
82. B. Drost, et al., *Introducing mvtec itodd-a dataset for 3d object recognition in industry*, in Proceedings of the IEEE International Conference on Computer Vision Workshops, 2017. pp. 2200–2208
83. G. Ros et al., The SYNTHIA dataset: A large collection of synthetic images for semantic segmentation of urban scenes. Proc. IEEE Conf. Comput. Vis. Pattern Recognit., 3234–3243 (2016)
84. J. Long, et al., *Fully Convolutional Networks for Semantic Segmentation.* In IEEE Conference on Computer Vision and Pattern Recognition (CVPR'15), 2015. 10.1109/CVPR.2015.7298965
85. M.A. Bolstad, *Large-Scale Cinematic Visualization Using Universal Scene Description.* IEEE 9th Symposium on Large Data Analysis and Visualization (LDAV'19), 2019. pp. 1–2
86. B. Calli, et al., *The ycb Object and Model Set: Towards Common Benchmarks for Manipulation Research.* International conference on advanced robotics (ICAR'15), 2015. pp. 510–517
87. J. Tremblay, et al., *Falling Things: A Synthetic Dataset for 3D Object Detection and Pose Estimation.* CVPR Workshop on Real World Challenges and New Benchmarks for Deep Learning in Robotic Vision, https://github.com/NVIDIA/Dataset_Synthesizer, 2018
88. T. Hodan, et al., *T-less: An rgb-d Dataset for 6d Pose Estimation of Textureless Objects.* IEEE Winter Conference on Applications of Computer Vision (WACV'17), 2017. pp. 880–888
89. F. Steinbrucker, et al., *Volumetric 3d Mapping in Real-Time on a cpu.* IEEE International Conference on Robotics and Automation (ICRA'14), 2014. pp. 2021–2028

Chapter 6
Creating SORDI: The Largest Synthetic Dataset for Industries

Smart robots in industrial settings and factories are increasingly performing differ-ent kinds of tasks, in coordination and collaboration with human workers, in order to improve industrial processes and boost manufacturing performance. As described in the previous chapters of this book, human-machine collaboration through robots and software agents is at the center of the Industry 5.0, where machines help humans extend their craftsmanship and analytical skills to deliver higher quality products and services. Nonetheless, providing seamless collaboration between humans and machines is not an easy task, and requires a huge amount of planning, training, and testing, to attain a useful and safe collaboration [1]. A typical scenario is that of Idealworks's iw.hub, an Autonomous Mobile Robot (AMR) designed to allow driv-erless transport of materials and goods within warehouses and factory floors. It moves around using a battery of cameras and sensors for navigation and obstacle avoidance. To achieve these tasks, Computer Vision (CV) allows the iw.hub robots to observe, monitor, recognize, and meaningfully understand their surroundings, e.g. detecting and recognizing specific objects in a scene, in order to perform spe-cific actions, collision avoidance, and path rerouting accordingly. These CV tasks can be achieved after training deep learning (DL) models on large annotated datas-ets. In industrial settings, obtaining and labelling such datasets is challenging because it is time-consuming, prone to human error, and limited by several privacy and security regulations. As a result, BMW Group and Idealworks have jointly developed a Synthetic Object Recognition Dataset for Industries (i.e., SORDI), the largest synthetic industrial dataset of its kind. Created using NVIDIA's Omniverse, SORDI consists of more than 100 industrial assets in 35 scenarios and more than 1,000,000 photo-realistic rendered images that are annotated with accurate pixel-level bounding boxes. For evaluation purposes, multiple object recognition models were trained with synthetic data to infer on real images captured inside a factory. Accuracy values higher than 80% were reported for most of the considered assets, highlighting the potential of the dataset and its practicality.

***

J. Nassif et al., *Synthetic Data*, https://doi.org/10.1007/978-3-031-47560-3_6

How is SORDI built? How are the material designed? How are the images and scenes generated? How is the dataset used for object recognition?

***

We attempt to answer these questions in this chapter of the book...

6.1 Universal Scene Description

The SORDI image dataset generation is inspired by Pixar Animation Studios' core graphics and rendering pipeline [2]. It allows rendering realistic and complex cinematic scenes using the Universal Scene Description (USD) framework. To put things into perspective, scientific visualization traditionally utilized simple rendering techniques such as Gouraud [3] or Phong [4] shaded polygons or ray sampled transfer functions for volume rendering [2], in order to transform and represent complex numerical data into images or animations. However, new and powerful hardware like NVIDIA's Jetson boards and GPUs, with their dedicated Isaak SDKs, have enabled the use of more computationally-heavy rendering techniques to allow for high interactivity with complex visual data. Recently, Pixar put forward its Universal Scene Description (USD) [5], an open source framework that serves as the core of Pixar Animation Studio's graphics and rendering pipeline, to allow for easy interchange of elemental assets, models (i.e., groups of assets), or animations. Different from other rendering packages, USD enables the grouping and organization of any large number of assets into virtual sets, scenes, and shots. These can be easily exchanged between applications, using a central API to edit them as overrides. USD provides a rich toolset for reading, writing, editing, and speedily previewing 3D geometry and shading. USD also provides a high-performance OpenGL[1] based application to preview the generated files [6], which can be easily read and edited in their raw ASCII for binary formats (cf. Fig. 6.1). USD has been widely adopted across a wide variety of digital content creation tools (including Maya, Houdini, and Katana with Blender and 3DMax support under development) and game engines such as Unity, Unreal Engine, and NVIDIA's Omniverse. As such, it makes an attractive target to bridge the gap between scientific data and tools for cinematic visualization. Also, since USD's core scene graph and composition engine are not tied to any specific visualization format, USD can be extended in a maintainable way to encode and compose data in different domains [2], namely in different industrial settings such as BMW Group's logistic, supply chain, and manufacturing scenarios. When producing SORDI, separate teams at BMW Group and Idealworks worked on specific areas and scenarios, where every team was responsible for delivering, updating, maintaining, and exchanging their complex assets or materials (cf. Fig. 6.2). These assets were then assembled to construct more complex parent scenes that matched their real environment counterparts (cf. Fig. 6.3).

---

[1] https://www.opengl.org/

```
det Morro "World
{
   def Mesh "mesh_0"
   {
      float3[] extent.timesamples = (
      1: [(-0.5, -0.5, -0.5), (0.5, 0.5, 0.5)],
      int[] facellertexCounts.timesamples = {
      1: [4, 4, 4, 4, 4, 4],
      int[] facevertexlndices.timesamples = f
      1: [1, 5, 4, 0, 2, 6, 5, 1, 3, 7, 6,2, 0, 4, 7, 3, 2,1,
      0, 3,5, 6,7, 4],
      point3f[] points.timesamples =(
      1: [(-0.5, -0.5, -0.5), (0.5, -0.5,-0.5),(0.5, -0.5,
      0.5), (-0.5,-0.5,0.5),
      (-0.5, 0.5, -0.5), (0.5, 0.5,-0.5),(0.5, 0.5,0.5), (-
      0.5, 0.5, 0.5)],
      color3f[] primvars:displayColor (
      interpolation = ''vertex"
      color3f[] primvars:displayColor.timesamples =
      1: [(1, 0, 0), (1, 0.5, 0), (1, 1, 0), (0, 1, 0), (0, 1,
      1), (0, 0, 1), (1, 0, 1), (1, 0.5, 1)],
      }
   }
}
```

Fig. 6.1 USD sample ASCII file (left) and the geometry it describes (right) [2]

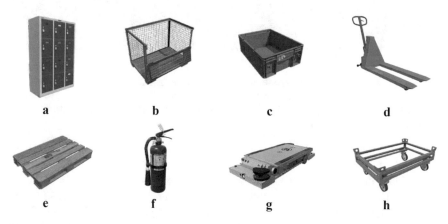

Fig. 6.2 Sample SORDI assets: (**a**) Cabinet (**b**) Stillage (**c**) KLT (Kleinladungstrager) Box. (**d**) Jack (**e**) Pallet (**f**) Fire Extinguisher (**g**) AMR (Autonomous Mobile Robot), and (**e**) Dolly

6.2   3D Mesh Modeling

As mentioned previous, USD is an open-source framework with wide industry adoption. It is supported by companies such as Autodesk, Apple, Blender, and NVIDIA. More specifically, NVIDIA's Omniverse enables live collaborations between different applications, e.g. 3DsMax, Unreal, etc., that support exporting renders in the USD format. Therefore, as an advantage, it is not necessary to re-model existing 3D assets, e.g., whole factories, machines, and robots that have already been modeled in one of the aforementioned digital content creation software. In the case of SORDI, various BMW Group and Idealworks teams used 3D mesh modeling to capture BMW industrial assets from different points of view and

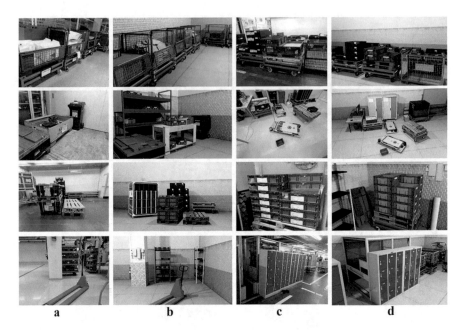

Fig. 6.3 Complex scenes content generation based on real industrial scenarios, comparing the real photo captures (**a** and **c**) with their SORDI synthetic replicas in NVIDIA's Omniverse (**b** and **d**). (**a**). Real capture. (**b**). Synthetic replicas based on images from *a*. (**c**). Real captures. (**d**). Synthetic replicas based on images from *c*

re-modeled them using Blender. 3D scanning is performed to collect data on the shape and appearance of every asset. Different from regular cameras, 3D scanners collect distance information about the surfaces within its field of view, describing the distance to the surface at each pixel in the produced image. The data is collected in the form of a polygon mesh point cloud, and is processed to extrapolate the shape of the object and construct a digital 3D model of it. The 3D model is then processed for graphics rendering as described in the following section. The teams at BMW Group and Idealworks used 3D laser scanning, which sends out laser beams that reflect back when they hit objects, at which point the distance the laser beam has travelled is measured [7]. Measurements are stored as points with *xyz*-coordinates relative to the scanner-position, creating a cloud of points that visualizes the scanned environment. The point clouds' density depends on the laser scanners' performance and the set resolution [8]. Several scans of large areas and several sides of every asset were captured separately, and then merged together to form a large point cloud (cf. Fig. 6.4). Computer-generated objects are also imported into the point cloud, to help realize more realistic and complete BMW virtual environments.

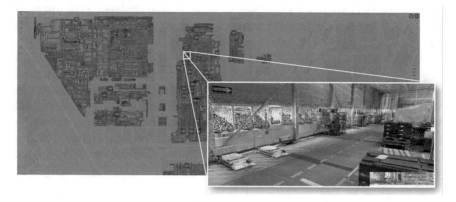

Fig. 6.4 Virtual factory layout of BMW Group's Regensburg plant, using point cloud-based 3D object representations

6.3 Material Design

Inspired by the real asset surfaces and textures, the 3D modeling team optimized manual photorealistic parametric materials using Substance 3D Suite.[2] The Substance 3D Modeler allows creating 3D models based on vector graphics and mathematical shapes. The Substance 3D Sampler allows extracting textures from real-life pictures and transforming them into photorealistic materials, 3D objects, and rendering environments. The Substance 3D Designer allows manipulating the 3D model and its superimposed textures to create seamless materials, patterns, filters, and environment lights with a huge battery of variations according to the designer's needs. Materials (or textures) are the images that are applied to the different surfaces of a 3D object to give the object a more realistic look and feel [9]. Applying materials on a 3D object is a step referred to as texturing, which is a central step toward 3D image rendering. It allows providing each 3D surface with the color and pattern properties needed to visually describe the material the object is made of [10]. In other words, texturing a 3D object makes it look like it's made out of plastic, metal, wood, or any other material. Substance 3D also provides a large library of assets and textures, including around 9000 materials with visually customizable parameters for different purposes and styles [11]. Here, we distinguish between two types of materials: (i) fully procedural, and (ii) scan-based. On the one hand, fully procedural materials are designed exclusively virtually created with procedural tools like Substance 3D Designer [9]. They are fully parametric and offer the highest level of flexibility as everything in them is potentially customizable. These assets have the most exposed parameters that can be controlled by the user. Their file sizes are usually limited to a few kilobytes and allow to generate high-resolution renders given their mathematical nature. On the other hand, scan-

[2] https://www.substance3d.com/

based materials consists of textures that have been essentially captured by a material scanner or camera. This provides realism in capturing the details of the real-world material, providing an authentic reproduction of reality. Nonetheless, scan-based materials are usually more constrained than their procedural counterparts, since they are bounded by realism. In the case of SORDI, the pros of both procedural and scan-based materials are leveraged by using the Image-to-Material AI tools provided by Substance Alchemist [12], in order synthesize and generate real-life inspired materials. A battery of parameters describing color, roughness, metallicity, patterns, and surface relief, are fine-tuned to generate realistic industrial factory assets, while allowing controlled parameter variation to create randomized assets and environments.

6.4   Scene Rendering and Data Cleaning

As discussed in Chap. 3, different rendering techniques have been developed in the past years, including rasterization, ray casting, and more recently ray tracing (cf. Chap. 3, Sect. 3.4). In the case of SORDI, assets are rendered using ray tracing [13], due to its higher image fidelity compared with its counterparts. In practice, legacy global illumination algorithms are not suited for industrial locations and factories, since objects in factories and warehouses are typically lit by many fluorescent tube lights and there is much indirect lighting due to bright walls. In contrast, ray tracing allows diffused lighting, by tracing a path from an imaginary eye through each pixel in a virtual scene, and computing the color of each individual object visible through it. Every ray is evaluated for intersection with other objects in the scene, allowing the algorithm to estimate the incoming light at the point of intersection with the nearest object, examine the material (texture) properties of the object, and aggregate these two pieces of information together to compute the resulting color of the pixel. While it may seem awkward to send light rays *away* from the camera, rather than *into* it as actual light does in the real world, yet doing so is much more efficient computationally. Given that the great majority of light rays from a certain light source do not make it directly into the viewer's eye, a "forward" simulation can squander a huge amount of computation on light paths that are seldom recorded [14]. Hence, the idea behind ray tracing is to consider that a certain light ray intersects the view frame, and after a maximum number of reflections or a ray traveling a certain distance without intersection, the ray terminates its travel path and the pixel's value is updated [14]. Nonetheless, rendering remains a computationally intensive task, especially when dealing with complex scenes and large numbers of assets. A possible solution is to pre-render the images offline and then using them when needed. Yet real-time rendering is sometimes required for mixed reality simulation environments such as industrial simulations that need to dynamically create scenes and adapt to the workers' and robots' behaviors on-the-fly. In such applications, 3D hardware accelerators can improve real-time rendering by using the graphics processor on the video card (GPU) instead of consuming valued CPU resources [15]. In the case of SORDI, two 24 GB RTX 3090 multi-GPU systems were used to execute the rendering on NVIDIA's Omniverse platform [16].

To build virtual factory layouts using SORDI, synthetic assets were added to synthetically generated scenes by mapping their usage in a factory or in a warehouse. Some scenes contain single assets (cf. Fig. 6.5, 1–4) or several assets (cf. Fig. 6.5, 5–8). Each asset might also have either one instance (cf. Fig. 6.5, a1, b2, c2, b3, c3, a4) or multiple instances (cf. Fig. 6.5, b1, c1, d1) in a scene. More assets (e.g., whiteboards, industrial racks, office desks, electrical boxes, etc.) are added to the scenes to ensure randomness while capturing images from different angles.

Fig. 6.5 Synthetic images containing various scenes of different industrial scenarios

While one of the main advantages of synthetic image generation is the ability to automatically obtain accurate annotations, yet these annotations are sometimes over-accurate to the pixel level where each image pixel is annotated [16]. As a result, small bounding boxes for far away objects, small objects, or hidden objects are generated on top of the rendered scenes. Yet this produced another problem: reducing the training accuracy of the object recognition model, especially when bounding boxes encompass completely hidden objects or extremely small objects. As a result, labels of completely hidden objects or barely visible objects were removed from SORDI, preserving only those clearly distinguishable objects. Additionally, generating larger numbers of synthetic images in a small random space, eventually led to similarities among the synthetic images. Hence, an algorithm was implemented to (i) hash each image into an 8×8 monochrome thumbnail, (ii) measure the similarity between two hashes based on the Manhattan distance between the image's representative low-level features, and (iii) remove the images whose distance is below a predefined threshold [16].

6.5   Dataset Description

SORDI currently consists of more than 100 industrial assets (cf. Fig. 6.2) organized in a layered taxonomy,[3] including parent-child inheritance (IsA) relationships connecting related asset classes. It currently includes more than 1,000,000 images, automatically captured in 35 different scenes as presented in Fig. 6.5. The first scenes contain single asset scenarios, with a possibility of having multiple instances per image. The rest of the scenes are associated with factory and warehouse similar representations with numerous assets and instances. The camera position and rotation are randomized for every image, rendered in 720p resolution. Also, we apply transformation domain randomization by varying x and y axis position and z axis rotation for some of the annotated assets. As a result, SORDI's one million synthetically generated images have more than 6.5 million instances where each image contains on average 2 assets and 7 instances [16]. Note that the assets are not distributed in an equal manner due to their different sizes and usage. As shown in Fig. 6.6.a, the most common assets are pallets (21.25%), dollies (19.50%), and KLT boxes (16.01%), since they act as containers or holders in a factory or in a warehouse. Numerous KLT boxes can be found in a single scene, due to their small size and usage, hence the high number of instances in Fig. 6.6b. Concerning the other assets (e.g., STR in Fig. 6.5a7), each one of them is available in around 8.65% of the complete dataset. Besides, 45.12% of the dataset consists of single asset captures. It is possible to have multiple instances of the same asset in the same image (cf. Table 6.1). Yet, less than 6.29% of the images include more than 5 assets each.

---

[3] https://sordi.ai/tree

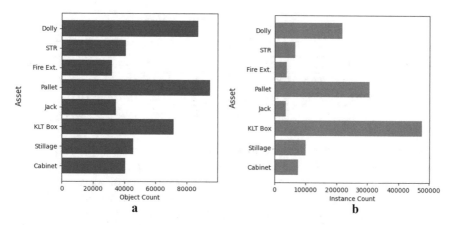

Fig. 6.6 Snapshots of SORDI dataset object and instance statistics [16]. (**a**). Object statistics. (**b**). Instance statistics

Table 6.1 Percentage of different asset occurrences in a single capture [16]

Occurrence	1	2	3	4	5	6	7	8
Percentage	45.12%	32.14%	14.72%	10.72%	4.25%	0.53%	0.43%	1.08%

6.6   Usage for Object Recognition

As mentioned previously, the SORDI dataset is primarily designed to train deep learning (DL) computer vision (CV) models to perform object recognition on real images. Early experiments focused on single class object recognition models. For each object recognition model, 3000 labelled SORDI images were randomly selecting as a training set, and 300 real images were captured in a BMW plant and annotated manually to form the test set. The manually annotated bounding boxes were considered as the ground truth labels throughout the evaluation process. Transfer learning was employed based on the following DL object recognition model architectures: FRCNN Resnet-50 [17], FRCNN Resnet-101 [18], SSD Inception and SSD Mobilenet [19]. All pre-trained model weights were based on the COCO dataset [20]. Experiments were conducted on Tesla V100-SMX2-16GB GPUs, with the training and evaluation codes made available online.[4] The DL models were evaluated by reporting the percentage of correct predictions over the entire test set (cf. Eq. 6.1).

$$Accyracy = \frac{TP + TN}{TP + TN + FP + FN} \in [0,1] \tag{6.1}$$

---

[4] https://github.com/BMW-InnovationLab/BMW-TensorFlow-Training-GUI

Table 6.2 Accuracy levels of different object detection models [1]

		Assets							
		Cabinet	Stillage	KLT box	Jack	Pallet	Fire Ext.	STR	Dolly
FRCNN	Resnet-50	0.4948	**0.887**	0.539	0.966	0.42	0.905	**0.795**	**0.806**
	Resnet-101	0.7784	0.878	**0.562**	1	0.495	1	0.787	0.74
SSD	Inception	**0.9585**	0.582	0.285	0.867	0.273	0.714	0.794	0.5
	Mobilenet	0.9482	0.763	0.076	1	**0.5**	0.84	0.603	0.529

where *TP* and *TN* represents the number of correct predictions (i.e., true positives and true negatives), while FP and FN represent the number of incorrect predictions (false positives and false negatives). Accuracy is maximized with a maximum number of correct predictions and a minimum number of incorrect predictions. Results were promising, achieving 100% accuracy for the jack and the fire extinguisher assets using the FRCNN Resnet-101 architecture, and > 95% accuracy circa for the cabinet asset using both SSD-based models (cf. Table 6.2). FRCNN Resnet-50 performed best when detecting stillages, STRs, and dollies achieving accuracies of 88.70, 79.50, and 80.60% respectively. Nonetheless, lower accuracies were also achieved with other assets like pallet (50.00%) and KLT box (56.20%). One possible explanation is that the pallet and KTL box assets (cf. Fig. 6.2) are not characterized by a unique texture, shape, dimension, or location, and they are usually either stacked or sided next to each other in factories and in warehouses (cf. Fig. 6.3). As a result, the object recognition models did not perform well when a group of KLT boxes or pallets were indistinctively grouped together or shelved on top of each other without any clear separation [16]. This would require improvement by producing more scenes with domain randomizations to further generalize the dataset. More object classes also need to be added, to distinguish between different kinds of pallets and KLTs, in order to improve object recognition performance while describing more realistic scenarios. Generative networks for image generation and data augmentation can be utilized to decrease the reality gap between real and synthetic images. To support the latter extensions and updates, an automated pipeline for synthetic image generation and rendering is being developed as part of the ongoing improvements on SORDI, which we further describe in the remaining chapters of this book.

6.7   BMW Group GitHub

As part of its SORDI and digital twin initiatives, the BMW Group TechOffice in Munich maintains a battery of github repositories describing and publishing the office's activities in collaboration with its academic and industry partners, namely Idealworks. The repositories describe the team's activities around physics-based synthetic data generation to produce the SORDI assets [21], SORDI assets evaluation interface [22], synthetic data cleaning [23] and selection [24], visual object labelling [25], object recognition deep model training [26], and visual data anonymization [27], among others.

Fig. 6.7 Sample snapshot images from the TechOffice digital twin including empty areas or irrelevant objects that are not useful for certain image classification or object recognition tasks

For instance, the synthetic data generation repository [21] provides an API that allows generating physics-based synthetic datasets the likes of SORDI, with just a few clicks specifying user input preferences. The AI evaluation repository [22] allows evaluating a trained CV model and acquiring general information and evaluation metrics with little configuration. The user provides a labeled dataset that is used as ground truth to assess the model, and an inference API that is used to infer on the selected dataset. Users can utilize one of the inference APIs provided by the TechOffice [28]. The BMW Labeltool Lite [25] can be used to label the dataset, where the images and their labels are used directly for evaluation. The evaluation GUI supports both object recognition and image classification. The synthetic data cleaning repository [23] allows cleaning image annotations with little to no configuration from the end user. It takes into consideration depth information, occlusions, and small bounding box inconsistencies that can be automatically corrected [29]. The image data selection repository [24] provides functions to select relevant images in a dataset. Selection is based on disregarding empty areas and irrelevant objects in the images (cf. Fig. 6.7). The TensorFlow training GUI repository [26] allows users to easily train a state-of-the-art DL model with little to no configuration needed. Users provide their labeled datasets and start the training process right away, while monitoring the process with TensorBoard. Users can also test their models with the TechOffice built-in inference REST API. The data cleaning repository [23] provides a dynamic approach to assess image quality for cleaning, based on dedicated image quality metrics including: (i) distribution threshold and (ii) density threshold. The motivation is to achieve high asset distribution so the image is not empty, and to have low asset density in order to avoid unwanted overlappings. Given the importance of data privacy and individuals' anonymity in industrial settings, the BMW anonymization tool [27] allows localizing and obfuscating (i.e., hiding) sensitive information in images and videos in order to preserve the individuals' anonymity. The tool is agnostic in terms of localization techniques, and supports semantic segmentation [30] and object recognition [31]. We further describe some of the latter tools and APIs in the last chapter of this book (cf. Chap. 8) highlighting the latest improvements and additions to SORDI.

References

1. A. Pinker, M. Pruglmeier, *Innovations in Logistics* (Huss, 2021), p. 192
2. M. Bolstad, *Large-Scale Cinematic Visualization Using Universal Scene Description*. IEEE 9th Symposium on Large Data Analysis and Visualization (LDAV'19), 2019. pp. 1–2
3. H. Gouraud, Continuous shading of curved surfaces. IEEE Trans. Comput. **C-20**(6), 623–629 (1971)
4. A. Watt, M. Watt, *Advanced Animation and Rendering Techniques: Theory and Practice* (Addison-Wesley Professional, Boston, 1992), pp. 21–26
5. Studios, P.A., *Introduction to USD*, 2019. https://www.pixar.com/usd
6. OpenUSD, *Utility Classes for OpenGL*, 2023. https://openusd.org/dev/api/usd_vol_page_front.html
7. L. Klein et al., Imaged-based verification of Asbuilt documentation of operational building. Autom. Constr. **21**(I), 161–171 (2012)
8. M. Dasso et al., The use of terrestrial LiDAR technology in forest science: Application fields, benefits and challenges. Ann. For. Sci. **68**(5), 959–974 (2011)
9. Adobe, *Limitless 3D Assets Creation*, 2023. https://www.adobe.com/products/substance3d-designer.html
10. Adobe, *Paint in 3D. In Real Time*, 2023. https://www.adobe.com/products/substance3d-painter.html
11. Adobe Substance 3D Content Team, *Parametric Materials: Back to the Source of Substance 3D Assets!* 2021. https://substance3d.adobe.com/magazine/parametric-materials-back-to-the-source-of-substance-3d-assets/
12. Adobe, *Digitize Your World with Substance Alchemist*, 2020. https://substance3d.adobe.com/magazine/digitize-your-world-with-substance-alchemist/
13. T. Viitanen et al., MergeTree: A fast hardware HLBVH constructor for animated ray tracing. ACM Trans. Graph. **36**(5), 1–169 (2017)
14. J. Peddie, *Ray Tracing: A Tool for All* (Springer, New York, 2019), pp. 1–358
15. A. Tewari et al., State of the art on neural rendering. Comput. Graph. Forum **39**(2), 701–727 (2020)
16. Akar C. Abou et al., *Synthetic Object Recognition Dataset for Industries*. International Conference on Graphics, Patterns and Images (SIBGRAPI'22), 2022. pp. 150–155
17. S. Ren et al., Faster R-CNN: Towards real-time object detection with region proposal networks. Adv. Neural Inf. Proces. Syst. **28**, 91–99 (2015)
18. Z. Wu et al., Wider or deeper: Revisiting the resnet model for visual recognition. Pattern Recogn. **90**, 119–133 (2019)
19. L. Barba-Guaman et al., Deep learning framework for vehicle and pedestrian detection in rural roads on an embedded GPU. Electronics **9**(4), 589 (2020)
20. H. Yu et al., *TensorFlow Model Garden*, 2020. https://github.com/tensorflow/models
21. BMW Group Innovation Lab, *BMW Physics Based Synthetic Data Generation*, 2023. https://github.com/BMW-InnovationLab/BMW-Physics-Based-Synthetic-Data-Generation
22. BMW Group Innovation Lab, *SORDI AI Evaluation GUI*, 2023. https://github.com/BMW-InnovationLab/SORDI-AI-Evaluation-GUI
23. BMW Group Innovation Lab, *BMW Synthetic Data Cleaning*, 2023. https://github.com/BMW-InnovationLab/BMW-Synthetic-data-cleaning
24. BMW Group Innovation Lab, *SORDI Image Data Selection*, 2023. https://github.com/BMW-InnovationLab/SORDI-Image-Data-Selection
25. BMW Innovation Lab, *BMW Labeltool Lite*, 2023. https://github.com/BMW-InnovationLab/BMW-Labeltool-Lite
26. BMW Innovation Lab, *BMW TensorFlow Training GUI*, 2023. https://github.com/BMW-InnovationLab/BMW-TensorFlow-Training-GUI
27. BMW Group Innovation Lab, *BMW Anonymization API*, 2023. https://github.com/BMW-InnovationLab/BMW-Anonymization-API

28. BMW Group TechOffice, *BMW TechOffice Munich*, 2023. https://github.com/BMW-InnovationLab/

29. M. Ayle, J. Tekli, J. El-Zini, B. El-Asmar, M. Awad, et al., Bar – A reinforcement learning agent for bounding-box automated refinement. Proc. AAAI Conf. Artif. Intell. **34**(03), 2561–2568 (2020)

30. BMW Group Innovation Lab, *BMW Semantic Segmentation Inference API*, 2023. https://github.com/BMW-InnovationLab/BMW-Semantic-Segmentation-Inference-API-GPU-CPU

31. BMW Group Innovation Lab, *BMW TensorFlow Inference API*, 2023. https://github.com/BMW-InnovationLab/BMW-TensorFlow-Inference-API-GPU

Chapter 7
Toward an Industrial Robot Gym

Factories represent industrial facilities made of several buildings occupied with different kinds of machines and robots, where workers produce or assemble manufactured items by operating different machines and working in tandem with the robots at the plant. Complex and expensive equipment, procedures, and transformation processes are often involved in manufacturing plants. Also, dangerous conditions may be required to operate different kinds of factories, including high voltage, high temperature, high pressure, high gas concentration, among other perilous situations. Hence, it is of central importance to understand how a manufacturing plant works, including its regular operations and the irregular situations that may arise, in order to better design and operate the plant with optimal safety, efficiency, and productivity conditions. This is where digital twins come into play and become of utmost importance. Modern factories are modeled as Cyber Physical Systems (CPS), including a smart physical system (i.e., the robotic machinery and infrastructure) controlled and monitored by an intelligent cyber system (i.e., the control software) [1]. Hardware and software components in a CPS are closely entwined, as a network of interacting robotic and sensor elements with physical input and output [2]. The digital twin represents an exact virtual copy of the factory or the CPS, including simulations of both their hardware and software components, allowing to simulate all their functionalities in a virtual simulation environment. Given the enormous, complex, and critical infrastructure involved in a factory, and given the difficulty of shutting a factory down or modifying its behavior to acquire a better understanding of its inner workings, the factory's digital twin becomes the next best alternative. It provides engineers with a virtual playground to test, change, revamp, reimagine, fast forward, and forecast the behavior of a plant in a completely virtualized environment, with zero impact or danger to its real counterpart. The digital twin can provide visualized information and interactive operations, while ensuring the security of the equipment and the safety of the human workers [3]. More specifically, the digital twin is viewed as a "robot gym", training virtual robotic twins using synthetic data in order to allow their physical counterparts to work and execute their tasks in the real world.

J. Nassif et al., *Synthetic Data*, https://doi.org/10.1007/978-3-031-47560-3_7

***

In this context, how do we build a digital twin and how do we make it work to benefit a modern factory? How can we provide quality assurance in the virtual world? How can we use synthetic data to train virtual robot twins?

***

We attempt to answer these questions in this chapter of the book.

7.1 Creating a Digital Mockup of the Factory

A digital twin is a virtual representation, also called a mirror, an avatar, or simply a twin of a physical system [4]. A digital twin consists of multiple integrated and interacting modules, including: (i) a system architecture identifying the main components making up the digital twin solution, (ii) a 3D model (i.e., a Virtual Factory Layout, or VFL) that is an accurate likeness of the physical system created with 3D modeling technologies, (iii) a mathematical model that governs the mechanisms, kinetics, and data properties of the physical system, and (iv) a rule-based model that allows user interactions and describes the dynamicity of the digital twin in correspondence with its physical counterpart. In the following, we further describe the latter components toward creating the digital twin of a factory.

7.1.1 System Architecture

Multiple system architectures can be considered when developing a digital twin solution. In the following, we describe a typical layered architecture that is suitable for the remote operations of the digital twin, which is especially useful in the context of digitalized factories. Remote operations mean the digital twin of the factory can be experienced by users remotely from the comfort of their offices or homes, through typical Web browsers or mobile applications, without having to connect directly to the computer server or to the cloud server machines hosting the digital twin. Figure 7.1 shows a sample design of a multi-layered architecture of a Web-enabled digital twin factory.

To enable remote operations of the digital twin, a layered architecture has been proposed regarding the perspective of use, deployment, and control [3]. The digital twin is run on dedicate cloud servers, where multiple twins can be designed to govern multiple aspects of the factory (e.g., multiple collaborating and integrated digital twins can be designed to govern each of the factory's main systems, including power supply, supply chain, assembly, etc., similarly to the collaboration and integration between their physical counterparts). Dedicated controllers ensure the data flow between the digital twins and their physical counterparts, which are essential

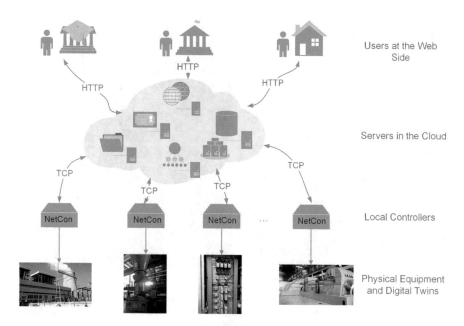

Fig. 7.1 Layered architecture of a Web/mobile-based digital twin plant [3]

to maintain synchronization among the virtual and physical systems. Users can access the digital twin from their personal computers, tablets, or mobile phones using Web/mobile renderings allowing to experience the digital twin with mainstream web browsers or mobile applications. Duplicate copies of the digital twin can be deployed on several servers to increase accessibility and allow different kinds or renderings (e.g., high-definition, lower-definition, simplified functionality, etc.) that are suitable for different kinds of users (e.g., engineers, managers, academics).

7.1.2   3D Model

3D models are used to create VFLs as virtual representations that are as similar as possible to their physical counterparts. In industrial settings, 3D laser scanning is usually utilized to capture the spatial properties of a factory. The 3D laser scanner measurements are stored as points with XYZ-coordinates relative to the scanner-position, creating a point cloud that visualizes the scanned environment [5]. To make the point cloud more easily editable, the list of points in 3D space can be transformed into a solid model using 3D modelling techniques such as voxelization and high-level 3D object representations. In this context, scene graphs have been increasingly used to encode and group geometric shapes into hierarchical structures, showing the benefits of 3D scene reconstruction using object detection in

Fig. 7.2 Real photo snapshots (**a**, **c**) versus photo realistic VFL snapshots (**b**, **d**) of BMW Group's Regensburg plant, mimicking the real plant using SORDI high-level 3D objects. (**a**). Real photo snapshots [6]. (**b**). Photo realistic 3D model renderings

point clouds (cf. Chap. 3). For instance, BMW Group and idealworks have adopted Pixar's Universal Scene Descriptor (USD) standard in generating their SORDI dataset 3D assets (cf. Fig. 7.2). USD is also adopted by NVIDIA in their virtualization engine Omniverse, which is used by BMW Group and idealworks to develop VFLs using SORDI. High-level representations introduces the concept of 3D objects and are not just a collection of points or voxels. This conceptual idea simplifies many otherwise complicated operations, including object and scene detection, recognition, translation, cropping, scaling, and rotation. In addition, high-level 3D representations are more easily rendered for visualization and manipulation on different platforms (e.g., personal computer, Web browser, mobile phone), as well as for physics and motion simulation, which are central operations for interacting with, manipulating, and controlling VFLs.

7.1.3   Mathematical Model

VFL models only resemble their physical model counterparts in appearance. However, to build a digital twin, the properties and dynamic behaviors of the physical factory need to be considered and simulated in the virtual environment, so that the behavior and animation of the 3D objects within the VFL resemble and match the dynamics of their real word mirrors. Here, the mathematical models differ in functionality, complexity and sophistication, from one object to the other, and from one factory to the other. For instance, the mathematical models of static physical objects like pallets and KLTs will basically govern their physical properties like 3D volume occupancy, weight, and material deterioration. The models of dynamic objects like transport robots and forklifts will govern their physical properties as well as their dynamic behaviors, including mobility, object detection, collision avoidance, and smart navigation within the plant floor, among other dynamic functionalities. The models of collections of objects or systems of objects, like the power supply system in a plant, or the conveyor belt system, consist of the integration of multiple underlying mathematical models governing the behavior of each object within the system, in addition to a system model that handles the coordination among the underlying object models and governs the properties and the behaviors of the system as a whole. Mathematical models usually include control algorithms involving dedicated procedures to ensure the physical and dynamic properties are achieved. For instance, techniques such as affine non-linear control [7], hierarchical model predictive control [8], and active disturbance rejection control [9] can be used to control large power plants [3] (cf. Fig. 7.3). In addition, state of the art control algorithms include machine learning models like deep learning and reinforcement learning to process big data including live data feeds and historical data records [10], in order to perform different operation scheduling, planning, maintenance, and forecasting tasks (e.g., forecasting the needed materials on the supply chain, or planning the optimal paths for a fleet of transport robots navigating the plant floor).

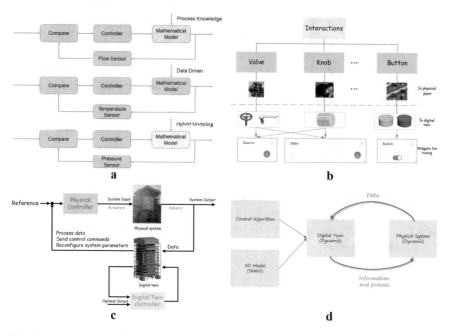

Fig. 7.3 Sample control, interaction, and rule models for a digital twin simulating a plant [3]. (**a**). Sample control algorithm diagram for a plant. (**b**). Motion control through user interaction. (**c**). Sample control block diagram between digital twin and physical system. (**d**). Sample rule model

Fig. 7.4 Using AR in BMW Group's iX5 Pilot Plant in Munich [11]

7.1.4 Rule Model

The rule model allows combining the control algorithm with the VFL to create a dynamic and interactive digital twin (cf. Fig. 7.3d). The interfacing and merging of the VFL with the control algorithm allows activating the virtual avatar that is the VFL by animating its behavior according to data coming from the physical system (e.g., the physical factory). In turn, the digital twin will also generate and process data according its mathematical model, where the data will be sent back to the physical system for feedback and adjustment if needed. Users can interact with both the physical system and its digital twin, by interfacing with the physical equipment in the real world and their virtual counterparts through the virtual renderings (cf. Fig. 7.3b). When interactions are done in the physical world, the data is fed back to the digital twin for real-time monitoring and control. The physical system can also send control commands, allowing to reconfigure system parameters as needed. As shown in Fig. 7.3c, the output of the physical system serves as one of the inputs of the digital twin, and the generated output of the digital twin controller acts as another input [3]. When interactions are done with the digital twin in the virtual environment, they are used for simulation, optimization, and forecasting. To achieve the perfect mapping with the physical system, the output of the digital twin should first converge to the optimal output, and then the optimal configuration can be fed to the physical system through the local controllers [3]. The digital twin configurations can also be fed back to the physical environment using Augmented Reality (AR) to allow for a double user interaction: (i) interacting with the real world through the user's physical presence in the factory, and (ii) interacting with the digital twin that is visually superimposed on top of the real world through AR visualizations and interactions with the virtual factory (cf. Fig. 7.4).

7.2 Augmented Digital Twin

Combined with the digital twin, Augmented Reality (AR) allows a higher level of interaction with both the virtual factory and its physical counterpart. It allows to visually superimpose parts of the digital twin on top of the real factory environment, allowing users to interact with the digital twin through its real word augmentation, and allowing for an improved interaction within the real factory environment by adding the superimposed visual information acquired from its digital counterpart [12]. In other words, AR allows to integrate the physical part and the virtual part of the factory in an intuitive, comprehensive, and adaptive way: different kinds of visualizations can be used at different times and in different places of the factory floor, to cater for different users' needs (e.g., electric engineers would prefer to see certain visualizations of the factory floor emphasizing electric circuitry and power flow, compared with manufacturing engineers or managers who would seek other kinds of logistic visualizations). Users can take advantage of the augmented data to perform more efficient decision-making and allow higher levels of machine control [13]. The combination of digital twin with AR, producing a so-called *augmented digital twin*, not only improves the efficiency and effectiveness of the manufacturing process, but also brings the user's experience in interacting with both the physical factory and its virtual counterpart to a completely different level.

7.2.1   Combining Physical and Virtual Parts

The digital twin provides real-time data and integrates it with historical data to provide more useful information to the system, allowing the system to self-evolve and improve itself. Depending on the properties of the digital twin data, its visualization using AR requires the following components [13]: physical part, virtual part, calibration process, augmented process, and control process. The physical part includes the physical objects in the real world, it can be a part, a product, a machine, or even the entire factory. All the data acquisition devices and sensors are also considered part of the physical part. The physical part makes up the founding layer, from which the other parts are aiming to analyze, utilize, and update the information [13, 14]. The virtual part consists of the VFL as well as all the real-time data collected from sensors, data acquisition devices, machines, and inputs from humans interacting with the digital twin. In addition, the historical data stored in the servers are also treated as part of the virtual part.

7.2.2   AR Calibration Process

In order to achieve an intuitive and accurate AR visualization of the digital twin, the 3D models in the VFL need to be faultlessly aligned with the physical part. This requires a delicate calibration process to accurately integrate the two parts together [15]. There are many calibration methods in AR, the most commonly used is the binary marker tracking method [16]. A binary marker is placed in the physical world where the AR devices can visualize and recognize it. A geometric object (usually a sphere or a cube) is designed in the virtual environment to align with the binary marker in the physical world. When the geometric object is perfectly aligned with the binary marker through the user's view, the calibration process is accurately done, allowing the virtual world to correctly aligning with the physical world where the VFL's 3D models are precisely overlaid onto their physical counterparts [16, 17]. More recently, markerless techniques have been introduced, namely Natural Feature Tracking (NFT) [18] and Simultaneous Localization And Mapping (SLAM) [19]. NFT-based solutions utilize computer vision models to detect representative points describing natural features in real-time video images [20]. Visual feature tracking algorithms are then used to produce accurate motion estimates and compute the virtual objects' pose accordingly [20]. An NFT-based solution has been recently developed to provide outdoor AR capability for users inside of a moving vehicle [21]. It continuously matches natural visual features from the camera against a prebuilt database of interior vehicle scenes, and combines pose estimation from both the vehicle navigation system and wearable sensors inside the vehicle. SLAM-based solutions consist in building a probabilistic feature map of the real environment in the form of a 3D point cloud, and then determines the AR navigation paths as a pre-scanning process. 3D objects constructed based on data from the sensors are compared with the predefined virtual models to estimate the virtual objects' poses [22].

7.2.3   AR Data Augmentation and Control

AR data augmentation consists in providing an adapted AR visualization of the digital twin data to users through AR devices. The AR device receives data from the virtual part and the corresponding calibration results, and then presents them to the users. This process does not merely display the entire virtual part directly onto the AR device. Different physical environments and objects, different input commands from users, and different AR devices entail different augmented processes to display different virtual objects on the AR devices [23]. For different machines and different users in the factory, the augmented process will display different information and virtual objects accordingly. With this filtering process, the digital twin data can be more understandable and useful to the users. Once the data is augmented and presented to the users, the control process allows the users to interact with both the physical part and the virtual part of the digital twin through AR. Users can control the physical part directly through the AR device by providing commands and inputs into the control process. This allows establishing a closed-loop control to improve and update of digital twin data, and continuously display the improved and updated data onto the AR devices [13, 15].

7.3   Virtual Quality Assurance

The quality of a digital twin can be evaluated according to three classification areas [24]: (i) Level of Development (LoD), (ii) Level of Accuracy (LoA), and (iii) Level of Recognizability (LoR, cf. Table 7.1).

7.3.1   Level of Development

LoD addresses the reliability of the digital twin models: which features they include, and what purposes they are designed for [25]. To ensure that a virtual object is used for its exact purpose, its purpose needs to be mapped to the functionalities that the object should perform. The description of a virtual object's functionalities can be organized under three main categories: (i) knowledge transfer and idea sharing, (ii) layout design and management, and (iii) simulations [26, 27]. The latter functionality groups include specific sub-areas. For instance, knowledge transfer can range from general static layout descriptions to specific dynamic object and machine designs [24]. Layout management can include new functionalities targeting new factories, or changes in the functionalities of existing factories that are adapted or repurposed to serve new factory designs. Simulations range over data/object/material flow, 2D movement, 3D movement, and robot simulation [28]. This variation in functionalities and purposes entails different requirements for defining them. For instance, most knowledge transfer functionalities aim at providing realistic virtual representations of the physical world as a means of communication, allowing to share knowledge about the physical system from and within the virtual world. This requires accurate and realistic VFLs that capture the look and feel of the real factory. Achieving layout management functionalities would require measurable metrics that users can employ to measure the distance between virtual objects in the VFL in order to allow their proper rearrangement [29]. This requires the objects' volume coverage areas and corners to be accurately defined allowing the user to easily choose areas and corners to measure between. Also, the user needs to move and displace objects in order to visualize how the new layout would look like. This means the virtual objects should be easily moveable in the virtual space [24]. Making an object moveable means it is created and located as its own separate part

Table 7.1   Features included in the classification of LoD, LoA, and LoR [24]

Level of development	Level of accuracy	Level of recognizability
Moveable objects	Very coarse	Object name
Measureable footprint	Coarse	2D area
Measureable 2D distances	Medium	3D block
Measureable 3D distances	Fine	Color
Object kinematics	Very fine	Shapes and features
Material flow		

inside the VFL so that it can be rearranged in relation with its surrounding objects. As for creating simulations of the objects or the whole production system, 3D motion studies are required to describe how the use of the LoDs can help define the requirements for specific simulation purposes [28]. For instance, 3D motion can be defined as a dynamic environment where an operator or a robot can be inserted into [24]. It requires 3D measurability of all virtual objects related to a potential operator or robot task, as well as surrounding objects that might interfere with the operations of the target object or system. 3D measurability allows measuring the required movement distances for objects and tasks, which can be compared against movement distances in the real world [26, 28].

7.3.2   Level of Accuracy

LoA emphases tolerances: how much a virtual object can differ within a certain tolerance, +/− levels of accuracy, compared with its real world counterpart [24]. Accuracy levels can be described using scalars ($\in [0, 100]\%$) or quantified into crisp linguistic categories (e.g., coarse, medium, fine) where the level of granularity depends on the users' preference and application scenarios. When performing layout management for instance, users need to know the accuracy of the outer dimensions of the objects they wish to align, otherwise users will not know what the VFL is valid for [28]. When simulating 3D robot movements for instance, a certain level of accuracy is required to make correct assumptions of the time required for the movements. Lower accuracy levels would entail lesser certainty levels in the robot's movements, and thus lesser trustworthiness in the digital twin's ability to simulate the robot's behavior. Tolerance measures can be defined based on existing tools such as IS0-2768 and USIBD LoA [30], and they can be combined with manufacturing documentation tools such as surface profile tolerances [31].

7.3.3   Level of Recognizability

For a VFL to fulfil its purpose, it is a necessity that the user of the layout understands what it illustrates. Since the experience and knowledge of users differ based on their profiles and experiences, the requirements on the layouts would also differ accordingly. Production engineers well familiarized with the factory's facilities can easily recognize objects by quickly looking at them, while other persons less familiar with the factory (e.g., engineers or technicians working in different fields, business personnel, consultants) may require more time or additional shapes or features, and sometimes even descriptive texts to recognize the objects [24]. LoR allows to support users, by helping VFL modelers with visual aids from the real world, allowing them to reconstruct the virtual objects with as much authenticity as possible, and to show the VFL end-users what kind of visual recognizability they can expect from the model [29]. The LoR features in Table 7.1 are pieces of information that can make an object more recognizable. Reference models of frequently occurring objects such as material racks, pallets, KLTs, etc., could be described by a couple of pictures and an added text describing how the model relates to the chosen features. This will provide both the project initiator and the VFL modeler with what to expect in terms of recognizability of an object [24].

Fig. 7.5 shows individual objects and groups of objects represented as block and shape models used for layout management. Fig. 7.5a, b represent the iw.hub Autonomous Mobile Robot (AMR) represented as a colored 3D block and as a group of specific shapes resembling the real robot respectively. In a transport navigation scenario, the block representation in Fig. 7.5a would be enough to define and recognize the robot, and its distinctive color would allow distinguishing it from other moving robots and mapping it with its source and destination locations by color coding them with the same AMR color in the VFL. The representation in Fig. 7.5b provides a realistic representation of the AMR allowing its recognition in a full-fledged realistic rendering of the virtual factory plant. The example in Fig. 7.5c provides a 3D block representation of the factory plant, at different layers of detail, which is suited for a user well familiarized with the factory layout and who wants to make changes to it (e.g., to improve it, to incorporate a new compartment into it, to define a new supply chain pipeline that feeds into it, etc.).

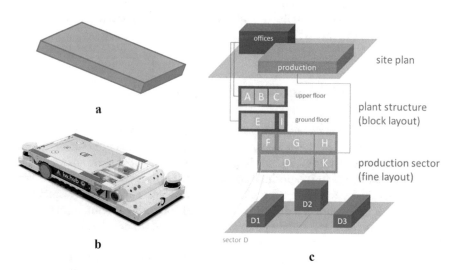

Fig. 7.5 Examples of LoR 3D block (**a**, **c**) and object shape (**b**) features. (**a**). 3D block model with blue color. (**b**). Model including specific shapes and features to describe the iw.hub robot. (**c**). Sample VFL for layout management [32]

7.3.4   Combining all Three Areas

While the individual classification areas are important in their own regard, yet each of them separately says little about the overall reliability and quality assurance of the digital twin solution. All three classification areas need to be combined in order to acquire a better assessment of what the digital twin can be used for and what is required by its users [24]. Without evaluating the recognizability of a virtual object, it is difficult to determine what is required by the user to understand the object. Also, the accuracy levels of the object and environment representations are required to assess the validity of the measures obtained from the VFL. As for its development, if the VFL does not have the features needed to fulfil the purpose of the digital twin, the latter's purpose will not be achieved. Combining the three classification areas can be defined once in a digital twin project, i.e., for the whole virtual factory, or different combinations can be considered for separate parts or separate objects of the virtual factory. Different combinations can be used for different purposes in the digital twin. For instance, 3D motion studies might be performed on an object or on a production line, whereas the rest of the VFL is only used for visual layout purposes. It could therefore be suitable to have general project requirements for a layout management task (cf. Fig. 7.5c) whereas the specific objects or lines have requirements combined with dedicated 3D motion studies. Nevertheless, depending on the amount of different applications of the digital twin, it might be wise to limit each project to a manageable amount of variations. Otherwise, the time gained from not overdeveloping the models can be lost in additional time for defining and keeping track of all the requirements and all the combination variations [24].

7.4   The Case of Iw.hub – BMW Group's AMR

Iw.hub is BMW Group's next generation AMR, aiming to increase the flexibility and efficiency of intra-logistic processes in the manufacturing plant. Developed by Idealworks, iw.hub is powered by state-of-the-art software and hardware that can plan the optimal route to efficiently execute material flows, while dynamically avoiding obstacles and other vehicles and safely operating around humans. Iw.hub is a digital twin enabled and trained robot, allowing for adaptive navigation and training in the virtual world, to efficiently and effectively anticipate, recognize, and adjust to changes in the real environment accordingly. For instance, changing a supply line or updating the location of a materials' depo position in the real plant will be anticipated in the digital twin, where the iw.hub is retrained to consider the new locations and constraints, before deploying it in the real factory.

7.4.1   Difficulties in Manufacturing Environments

The intricacy of manufacturing environments is chiefly due to their heterogeneity: they include stationary and moving objects, humans and machines, as well as autonomous and non-autonomous vehicles. For AMRs, things are even more complicated due to the difficulty of establishing predefined paths or a traffic rulebook that regulates the traffic within the factory. Although existing AMRs are usually equipped with the needed hardware and software to operate safely without human supervision, nonetheless, their behavior typically demonstrates a lack of smoothness due to their shortness in cognition abilities [33]. Figure 7.6 depicts some operational drawbacks encountered on a shop floor case study reported from [33]. In the convoy driving situation in Fig. 7.6a for instance, a desired behavior for the rear robot is to mimic the behavioral pattern of the front robot, without attempting to overtake, since the latter has an anticipated insight and thus has a more reliable judgment. In Fig. 7.6b, the field of view of the autonomous robot is lacking because of proximity. The autonomous robot needs to perceive the loaded forklift while approaching, and to deduce potential collision of loads in order to increase separation distance. In Fig. 7.6c, overtaking an obstacle on a two-way aisle might lead to a bottleneck with oncoming traffic. A more suited

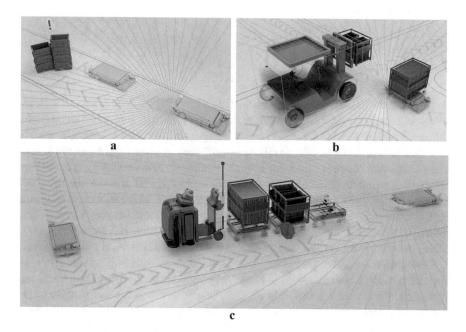

Fig. 7.6 Sample cases highlighting difficulties encountered with AMRs deployed in an automobile manufacturing plant [33]. (**a**). Convoy driving. (**b**). Hyperopic perception. (**c**). Oncoming traffic with long obstacle

behavior in this use case is to avoid overtaking long obstacles. The above motivating use cases highlight the need to make AMRs more aware of their surrounding situation, to allow for improved and increased functional autonomy.

7.4.2   Improving Situational Awareness

To address the difficulties described above, researchers from BMW Group and Idealworkds have developed the first situational awareness framework titled AWARE [33], specifically designed to augment AMRs in automobile manufacturing plants with situational awareness capabilities. It consist of a Knowledge Base (KB) representation, a set of rules, and a decision module. The AMR's observations or data streams are represented using a timestamp-based temporal RDF KB [34]. The decision module uses a set of rules to reason over the observations and the priors in order to adapt the behavior of the robot. The framework can be easily extended to support other applications of autonomous vehicles by adapting the KB and adding more rules to cover the application domain. Most importantly, the proposed solution is designed to be integrated within BMW Group's digital twin framework, where the iw.hub is trained in the virtual plant, before it's deployed in the physical plant (cf. Fig. 7.9).

7.4.3   Knowledge Base

A snapshot of the AWARE KB is shown in Fig. 7.7 [33, 35]. It includes time-invariant instances, such as instances of class *Decision* (e.g., *pause* and *adjustSafetyRange*) and instances of class *Procedure* (e.g., such as object detection models like YOLO [36] or DetectNet [37]). Also, instances of class *OperationalArea* and class *ConstraintZone* are time-invariant. *OperationalArea* describes parts of the factory with a particular functionality such as aisles or drop-off areas, while the class *ConstraintZone* refers to delimited surfaces in the plant where specific behavioral regulations apply such as zones with limited speed or limited capacity zones. The AWARE KB also includes time-variant instances characterized by a timestamp, which describe processed data from different intrinsic and extrinsic sensor streams. Data extracted from the AMR's internal state and its surrounding neighborhood is inserted into the KB as instances of class *Observation*. An example of two instances from class *Observation* is shown in Fig. 7.7. One observation is concerned with one feature of interest only. Numerous observations can be associated with the same timestamp. If multiple features of interest appear concurrently, such as multiple detections occurring within a single frame, an observation is created for each of them independently [33, 35].

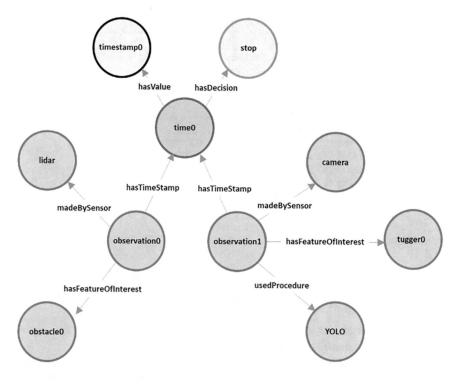

Fig. 7.7 Snapshot of the AWARE KB describing two robot observations

7.4.4   Assumptions and Rules

Different from road traffic, the functioning of AMRs in closed environments is not standardized by an established code of conduct. No published standard regulates traffic within the factory floor, such as intersections, right of way, or rules on when an AMR is supposed to yield way. Current AMRs usually perform the right of way allocation task following a *first come first served* approach, or following an *it fits I pass* policy. To improve on the latter, a set of behavioral rules were introduced to govern the behavior of AMRs deployed in a manufacturing environment [33], compared with the safety regulations put forward for driverless vehicles [38, 39]. The so-called AWARE rules are derived based on a set of assumptions to address the observed challenges and difficulties. We report some of the assumptions below [33] (cf. Table 7.2):

- *The behavioral rules are not considered as safety rules; instead, they are designed to ensure timely and orderly operations of the smart factory, where humans, manned vehicles, and AMRs function in alignment.*
- *Situational awareness is not a control system; instead, it is a guidance system facilitating the behavior of AMRs. In the absence of guidance, the AMR is supposed to proceed as indicated by its state machine.*

Table 7.2 Subset of AWARE rules written in SWIRL [33]

Convoy driving

```
Observation(?obs) ^ madeBySensor(?obs; camera)
^ hasF eatureOfInterest(?obs; ?obj) ^ AMR(?obj)
^ ObjectOfFocus(?obj) ^ TransitW ayObstacle(?obj)
^ hasTimeStamp(?obs; ?time) ^ TemporalEntity(?time)
! hasDecision(?time; stop)
```

Overtaking tugger train with oncoming traffic

```
Observation(?obs) ^ madeBySensor(?obs; camera)
^ hasF eatureOfInterest(?obs; ?obj) ^ Tugger(?obj)
^ ObjectOfFocus(?obj) ^ hasTimeStamp(?obs; ?time)
^ TemporalEntity(?time)
! hasDecision(?time; stop)
```

Hyperopic perception

```
isLoaded(ego, True) ^ Observation(?obs)
^ madeBySensor(?obs; camera) ^ hasF eatureOfInterest(?obs;
?obj)
^ Forklif t(?obj) ^ ObjectOfFocus(?obj)
^ hasTimeStamp(?obs; ?time) ^ TemporalEntity(?time)
! hasDecision(?time; increaseSafetyRange)
```

- *The AMR has always lower priority of way facing manned vehicles. This is due to the reduced agility of AMRs compared to manned vehicles.*
- *AMRs interact with each other following right of way rules similar to road traffic rules. That requires the ability for AMRs to recognize other AMRs and differentiate them from manned vehicles.*
- *All AMRs deployed in the same operations environment are expected to follow the same traffic rules.*
- *All AMRs deployed in the same operations environment are expected to have the same priors on the environment. Priors examples are intersections, driveway side, main and secondary aisles.*
- *AMRs cannot communicate between each others. To the best of our knowledge, no standard has been published to enforce lateral communication between AMRs. AWARE identifies unsolvable congestions and notifies the cloud master controller. Such standardized vehicle-to-vehicle communication is needed to guarantee complete autonomy.*

7.4.5　Framework Architecture

The iw.hub decision making system is built on top of the AWARE KB and is enriched by the behavioral rules discussed above. Reasoning over the statements in the KB, the iw.hub adapts its behavior and avoids different bottleneck situations by applying the inferred decisions such as *pause* or *increaseSafetyRange*. An architecture diagram of the framework is shown in Fig. 7.8. We briefly describe the main components below [33]:

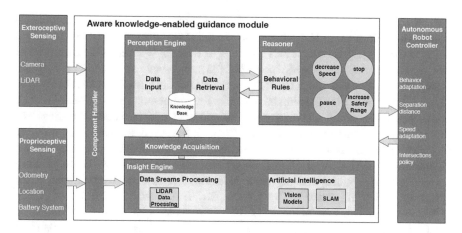

Fig. 7.8 Iw.hub's simplified AWARE guidance module architecture [33]

- **Component Handler**: It is the data extraction module, adapting the frame rate of data streams, and safeguarding the data alignment and its timestamps.
- **Insight Engine**: It is the data processing component, performing real-time data analysis. Data measurements extracted by *components handler* is structured according to the AWARE KB, before being processed by the *insight engine* (e.g., images captured by the camera are fed to a trained neural network for object detection).

- **Knowledge Acquisition**: It applies masks on the processed data to narrow down the insights to the area of focus, which varies with every sensor (e.g., for a camera sensor for example, the detected objects are filtered out following a trapezium of interest [40]).
- **Perception Engine**: It handles data input and data retrieval into and from the KB, by managing a time window of observations in memory.
- **Reasoner**: Also referred to as the *decision module*, it automatically checks the behavioral rules to trigger the ones that match the current instantiated state upon knowledge insertion. Depending on the observations in the target time window, the inferred guidance is sent to the control system to adapt the robot's behavior according to the perceived environment.

7.4.6   Using SORDI for Virtual Training in the Robot Gym

Reputable industry robotics solutions, including path planning and navigation, task planning, and manipulation problems, among others, utilize traditional optimization approaches. In contract, iw.hub's AWARE framework integrates semantic reasoning within the robot guidance system, to allow for increased behavior adaptability and avoiding bottlenecks when used on the factory floor. Nonetheless, for the designed

process to work properly, the robot needs to acquire advanced reasoning and understanding capabilities to allow for sophisticated knowledge acquisition and usage.
This is challenging for modalities like images where low-level pixels data need to
be interpreted into real world concepts [33]: to recognize encountered agents
through computer vision, a labeled dataset of all possible assets on the factory floor
is required, similarly to existing benchmarks for autonomous driving vehicles [41].
Such a dataset is rarely available and is difficult to acquire in industrial settings, for
many reasons discussed previously (cf. Chap. 5, Section 5.2), namely: (i) need for
manual human effort to capture, preprocess, annotate, and filter the images, which
is not always available, (ii) human object annotation is subjective and is prone to
errors, and iii) capturing images inside industrial locations and factories plants can
be difficult due to limited access for security. This is where SORDI and BMW
Group's digital twins come into play. SORDI provides a huge source of industrial
assets that can be leveraged to train iw.hub and other robots in the virtual plant, so
that it can perform its tasks correctly in the physical plant (cf. Fig. 7.9). Using
NVIDIA's Omniverse to build their digital twins, BMW Group and Idealworks continuously train their iw.hub virtual robots, allowing them to adapt to planned changes
in the environment, additions of new industrial assets, and adoption of new protocols, without affecting the work of the physical plant. As soon as a planned change
or process is announced or anticipated, it can be first simulated in the factory's digital twin, using the needed SORDI assets to populate the VFL. The iw.hub fleet then
is retrained accordingly, allowing them to update their AWARE knowledge reasoning accordingly. The object detection models are also trained in the virtual factory,
to recognize and detect SORDI assets encountered in manufacturing plants.
Consequently, the knowledge and reasoning gained from the virtual plant, i.e.,
which is used as the iw.hub's "robot gym", is processed and used to autonomously
run iw.hub in the physical plant.

Fig. 7.9 Retraining (**a**) and simulation testing (**b**) of a virtual iw.hub robot, before deploying the
retrained guidance system on the physical robot to execute in the physical plant. (**a**). Retraining of
a virtual iw.hub to pick up and drag a synthetic blue dolly from SORDI. (**b**). Simulation of the
retrained virtual iw.hub in the virtual factory populated with SORDI assets

7.5   Is the Digital Twin Worth It?

Despite the advantages of having a digital twin for the iw-hub AMR, and despite the improvements, enhancements, and excitement that the BMW Group digital twins are bringing to the automotive manufacturing environment, nonetheless, one always needs to take a step back and evaluate the pros and cons of every decision and every technical choice. This is part of the job of industry leaders: helping their teams avoid coming to agreement on technical or technological choices too quickly. Every now and then, leaders need to be slow in making decisions because of a need to think things over in a logical, analytical, and objective way, and offer measured, dispassionate and critical analyses regarding all aspects of the team's operations. This needs to happen at times of crucial decision making in order to consider competing proposals and suggest alternative ideas to any solution that is being used or being considered by their teams, including the development of digital twins in the context of BMW Group's digitalization strategy. This is also referred to as the *tenth man strategy*: every team of ten persons should have at least one person, i.e., the tenth man, doubting common ideas and suggesting alternatives. It is also referred to as the *devils' advocate* strategy which is adopted by the Vatican during the process of canonization (i.e., no matter how compelling the evidence is in support of canonization, there should always be a team of clergy taking the role of the devil's advocate and advocating against canonization). In this context, let's try to play the devil's advocate and advocate against digital twins. Here, we ask ourselves the following questions: *Should digital twins be developed for all industrial and manufacturing projects to achieve digitalization? When are digital twins not worthwhile?*

As mentioned in the previous chapters of this book, we have provided plenty of arguments, examples, and practical use case examples from our experience at BMW Group and Idealworks, and based on our expertise in the automotive logistics, robotics, and intelligent data processing sectors, in support for the development of and investment in digital twins as a means to achieve the digitalization of a factory. Playing the devil's advocate, we can safely state that not all projects are suited for digital twins. For retrofit projects which are simple enough or mature enough, it might be sufficient to use traditional design, implementation, and evaluation techniques in the physical world, without the need to produce a virtual replica of the project. The effort overhead needed to set-up, 3D scan, design, virtualize, implement, test, and post-work the digital twin can add more engineering hours than the traditional approach when it comes to small-scale and well known projects. Here, it is important to emphasize that producing a good quality and useful digital twin is a time consuming and resource consuming process: it takes a lot of time, a lot of manpower, and a lot of resources to build a full-fledged digital twin for a large and complex system like an automotive factory. A testimony to this are the huge investments and large-scale collaborations that BMW Group is establishing through its iFactory vision [42], in collaboration with tech giants like NVIDIA and Microsoft [43, 44]. In this context, we believe some sort of value-complexity onset should be established allowing industry leaders to make an informed decision of whether to pursue a digital twin initiative or not for a given project.

References

1. X. Yao et al., Smart manufacturing based on cyber-physical systems and beyond. J. Intell. Manuf. **30**(8), 2805–2817 (2019)
2. H. Kang et al., Smart manufacturing: Past research, present findings, and future directions. Int. J. Pr. Eng. Man. Green Technol. **3**(1), 111–128
3. Z. Lei et al., Toward a web-based digital twin thermal power plant. IEEE Trans. Industr. Inform. **18**(3), 1716–1725 (2022)
4. F. Tao et al., Digital twin-driven product design, manufacturing and service with big data. Int. J. Adv. Manuf. Technol. **94**, 3563–3576 (2018)
5. M. Dasso et al., The use of terrestrial LiDAR technology in forest science: Application fields, benefits and challenges. Ann. For. Sci. **68**(5), 959–974 (2011)
6. A. Padeanu, *2023 BMW X1 Enters Production At Regensburg Plant* (BMWBlog, 2022). https://www.bmwblog.com/2022/06/10/2023-bmw-x1-enters-production-regensburg-plant/
7. H. Zhou, C. Chen, J. Lai, X. Lu, Q. Deng, X. Gao, Z. Lei, Affine nonlinear control for an ultra-supercritical coal fired once-through boiler-turbine unit. Energy **153**, 638–649 (2018)
8. X. Kong et al., An effective nonlinear multivariable HMPC for USC power plant incorporating NFN-based modeling. IEEE Trans. Industr. Inform. **12**(2), 555–566 (2016)
9. Z. Wu et al., Gain scheduling design based on active disturbance rejection control for thermal power plant under full operating conditions. Energy **185**, 744–762 (2019)
10. S. Zhao et al., An overview of artificial intelligence applications for power electronics. IEEE Trans. Power Electron. **36**(4), 4633–4658 (2021)
11. A. Hemmerle, *Absolutely Real: Virtual and Augmented Reality Open New Avenues in the BMW Group Production System* (BMW PressClub Global, 2019). https://www.press.bmw-group.com/global/article/detail/T0294345EN/absolutely-real:-virtual-and-augmented-reality-open-new-avenues-in-the-bmw-group-production-system?language=en
12. C. Seidel, *Munich Pilot Plant: BMW Group Uses Augmented Reality in Prototyping* (BMW PressClubGlobal,2020).https://www.press.bmwgroup.com/global/article/detail/T0317125EN/munich-pilot-plant:-bmw-group-uses-augmented-reality-in-prototyping?language=en
13. Z. Zhu et al., *Visualization of the Digital Twin Data in Manufactoring by Using Augmented Reality.* 52th CIRP Conference on Manufactoring Systems, 2019. pp. 898–903
14. D. Castro et al., *Monitoring and Controlling Industrial Cyber-Physical Systems with Digital Twin and Augmented Reality.* IEEE International Conference on Consumer Electronics (ICCE'23), 2023. pp. 1–4
15. P. Bégout et al., Augmented reality authoring of digital twins: Design, implementation and evaluation in an industry 4.0 context. Fron. Virtual Real. **3** (2022). https://doi.org/10.3389/frvir.2022.918685
16. M. Kassim, A. Bakar, The design of augmented reality using unity 3D image marker detection for smart bus transportation. Int. J. Interact. Mobile Technol. **15**(17), 33 (2021)
17. C. Kim et al., *Marker Based Pedestrian Detection Using Augmented Reality.* International Conference on Advances in Image Processing (ICAIP'19), 2019. pp. 19–22
18. J. Lima et al., *Study About Natural Feature Tracking for Augmented Reality Applications on Mobile Devices.* Symposium on Virtual and Augmented Reality (SVR'15), 2015. pp. 7–14
19. W. Chen et al., SLAM overview: From single sensor to heterogeneous fusion. Remote Sens. **14**(23), 6033 (2022)
20. P. Fraga-Lamas et al., A review on industrial augmented reality systems for the industry 4.0 shipyard. IEEE Access **6**, 13358–13375 (2018)
21. Z. Zhu et al., *Head-Worn Markerless Augmented Reality Inside A Moving Vehicle.* IEEE Conference on Virtual Reality and 3D User Interfaces (VR) Workshops, 2022. pp. 680–681
22. H. Bae et al., Fast and scalable structure-from-motion based localization for high-precision mobile augmented reality systems. J. Mobile User Exp. **5**, 4 (2016)
23. Y. Yin et al., A state-of-the-art survey on augmented reality-assisted digital twin for futuristic human-centric industry transformation. Robot. Comput. Integr. Manuf. **81**, 102515 (2023)

24. A. Eriksson et al., Virtual factory layouts from 3D laser scanning – A novel framework to define solid model requirements. Proc. CIRP **76**, 36–41 (2018)
25. B. Forum, *Level of Development Specification – for Building Information Models* (BiM Forum, 2015), p. 195
26. A. Azevedo, A. Ahneida, Factory templates for digital factories framework. Robot. Comput. Integr. Manuf. **27**(4), 755–771 (2011)
27. M. Rohrer, *Seeing is Believing: The Importance of Visualization in Manufacturing Simulation.* Winter Simulation Conference (WSC'20), 2000. pp. 1211–1216
28. L. Giske et al., Visualization support for design of manufacturing systems and prototypes – Lessons learned from two case studies. Proc. CIRP **81**, 512–517 (2019)
29. Y. Fan et al., A digital-twin visualized architecture for flexible manufacturing system. J. Manuf. Syst. **60**, 176–201 (2021)
30. U.S. Institute of Building Documentation, *USIBD Level of Accuracy (LOA) Specification Guide Cl20,* 2016. Ver. 0.95 TM Guide
31. G. Henzold, *Geometrical Dimensioning and Tolerancing for Design, Manufacturing and Inspection*, 2nd edn. (Butterworth-Heinemann, Oxford, 2006), p. 416
32. T. Weber, *What is a Block Layout Useful for in Factory Planning?* (VisTable, 2023). https://www.vistable.com/blog/factory-layout-design/what-is-a-block-layout-useful-for-in-factory-planning/
33. B. El Asmar et al., *AWARE: A Situational Awareness Framework for Facilitating Adaptive Behavior of Autonomous Vehicles in Manufacturing.* International Semantic Web Conference (ISWC'20), 2020. pp. 651–666
34. C. Gutierrez et al., Introducing time into RDF. IEEE Trans. Knowl. Data Eng. **19**(2), 207–218 (2006)
35. B. El Asmar et al., *AWARE: An Ontology for Situational Awareness of Autonomous Vehicles in Manufacturing.* Proceedings of the 2021 Commonsense Knowledge Graph Workshop (CSKG'21)@AAAI'21, 2020
36. J. Redmon et al., *You Only Look Once: Unified, Real-Time Object Detection.* IEEE Conference on Computer Vision and Pattern Recognition (CVPR), 2016. pp. 779–788
37. A. Tao et al., *DetectNet: Deep Neural Network for Object Detection in DIGITS* (ParallelForall, 2016). https://developer.nvidia.com/blog/detectnet-deep-neural-network-object-detection-digits/
38. ANSI/ITSDF, *Safety Standard for Driverless, Automatic Guided.* B56.5, 2019
39. ISO, *Industrial Trucks – Safety Requirements and Verification – Part 4: Driverless Industrial Trucks and Their Systems.* 3691-4, 2020
40. R. Tapu et al., Deep-see: Joint object detection, tracking and recognition with application to visually impaired navigational assistance. Sensors **17**(11), 2473 (2017)
41. A. Geiger et al., Vision meets robotics: The KITTI dataset. In. J. Rob. Res. **32**(11), 1231–1237 (2013)
42. BMW Group News, *This is How Digital the BMW iFactory Is,* 2023. https://www.bmwgroup.com/en/news/general/2022/bmw-ifactory-digital.html
43. B. Caulfield, *NVIDIA, BMW Blend Reality, Virtual Worlds to Demonstrate Factory of the Future* (NVIDIA Blogs, 2021). https://blogs.nvidia.com/blog/2021/04/13/nvidia-bmw-factory-future/
44. J. Friedrich, *BMW Group's Innovative Edge Ecosystem Wins Award* (BMW Group Press Club, 2021). https://www.press.bmwgroup.com/global/article/detail/T0371373EN/bmw-group%E2%80%99s-innovative-edge-ecosystem-wins-award?language=en

Chapter 8
What Is Next with SORDI

While industry digitalization relies on a battery of AI-enabled digital transformation technologies, nonetheless, big data can be safely considered as the chief and indispensable enabler of modern industries. Smart industrial equipment and manufacturing robots are continuously ingesting and generating tons of data from and into the digitalized factory to perform all sorts of AI-enabled tasks. Visual data in the form of images and videos is required to train Deep Learning (DL) and Computer Vision (CV) models to perform object detection and recognition tasks for industrial robots on the factory floor and in the digital twin environment. Yet training DL and CV models require largescale industrial datasets, which are oftentimes unavailable due to the huge amount of time and resource needed to capture and manually label the data, as well as the limited access to manufacturing locations due to various security and privacy regulations. To address this problem, BMW Group, Microsoft, NVIDIA, and Idealworks have jointly developed the Synthetic Object Recognition Dataset for Industries (i.e., SORDI), the largest industrial dataset of its kind, in order to fuel BMW Group's "robot gym". SORDI allows training the robots' CV software in the virtual world before they execute in the real world, enabling Transfer Learning (TL) between the digital twin environment and the physical factory. The robot gym stands for the factory's digital twin, built using synthetic SORDI assets used to train the object recognition models. The physical plant is where robots execute the object recognition models already trained in the digital twin, in order to recognize objects from real images taken in the real factory. In the final chapter of this book, we describe the challenges and next development steps with SORDI to perform effective and efficient object recognition and TL. We discuss the integration of real-world assets and third party assets into SORDI to make it more realistic. We discuss the so-called reality gap between real and synthetic assets, and describe different solutions to breach the gap and bring synthetic images and environments closer to their real counterparts. We discuss the augmentation and partial automation of real data capture, and the integration of both real and synthetic datasets to perform DL model training. We also highlight the impact of using SORDI in smart

J. Nassif et al., *Synthetic Data*, https://doi.org/10.1007/978-3-031-47560-3_8

manufacturing and logistics applications in order to reduce the industry's carbon footprint, and allow for more sustainable production pipelines.

***

What is next for SORDI? How can we make SORDI more realistic, and its usage more practical? How can SORDI help the manufacturing process become more sustainable and help reduce its carbon footprint?

***

We attempt to answer these questions in the final chapter of this book...

8.1   Reality Gap

8.1.1   In the Beginning, There Was SORDI

Looking back at the field of CV for the past couple of years, one can clearly realize that DL models, namely Deep Convolutional Neural Networks (DCNNs) have been performing increasingly well on large public image datasets such as ImageNet [1] or MS COCO [2]. Having touched human-level performance in image classification, the main focus of CV research has gradually shifted toward object recognition, leading to the development of DL models such as faster Region-CNN (faster R-CNN) [3], Single Shot multi-box detector (SSD) [4], and You Only Look Once (YOLO) [5]. Even though these models are outperforming the traditional CV-based methods by a significant margin, their applications in industrial and robotic systems have many challenges. DL-based object recognition models require serious training on sufficiently large datasets, where the data is domain-specific depending on the industrial setting, and where the data is expertly labeled to reflect the target objects within the images. Nonetheless, as stated in previous chapters, collecting large industrial image datasets faces many hurdles, namely: (i) it requires a huge amount of resources and time to capture, pre-process, filter, and label the images, (ii) it is prone to human error and subjectivity during manual annotation, (iii) it is limited by several privacy and security regulations, as well as restricted access and maneuverability inside industrial locations and factories. To address the above challenges, BMW Group, Microsoft, NVIDIA, and Idealworks have jointly developed the Synthetic Object Recognition Dataset for Industries (i.e., SORDI), to allow TL between the virtual simulation environment and the real world. Sim-to-Real transfer is a special case of TL, where the source domain is the virtual simulation of the real world, while the target domain is the physical reality itself [6]. In this case, the virtual simulation consists of the factory's digital twin built using synthetic SORDI assets used to train the object recognition models. The real world consists of the physical factory floor where robots execute the object recognition models already trained in the digital twin, in order to recognize objects from real images taken in the real factory. In other words, SORDI is providing the data to run our virtual robot gym, to train the robots' CV software before they execute in the real world.

Hence the more realistic SORDI is, the better the quality of our robot gym, which begs the following questions: How can we make SORDI more realistic, and its usage more practical? How can we bridge the gap between the digital twin simulation training and the model's execution in the physical world, so that the training feels as realistic as possible?

In CV, the *reality gap*, also referred to as the *domain gap*, underlines the difference in performance between DL models trained on synthetic images versus models trained on real images. This stems from the complexity of the real-world: even if we consider controlled domains such as industrial and manufacturing environments, the real-world oftentimes includes a huge number of various combinations of environmental and behavioral factors, with prevalent events and rare events that are hard to reproduce in the synthesized simulation environment. The reality gap can be divided into two kings of gaps [7]: (i) visual, and (ii) content gap.

8.1.2 Visual Gap

The visual gap represents the appearance or perceptual difference between a synthetic image and a real image, including factors such as the image quality, colors, realism (namely with respect to the quality of the rendering system compared with a real camera sensor), in addition to the assets' shapes, materials, and details within the image. More specifically, the visual gap can be divided into [7]: (i) textures and materials, and (ii) lighting. A texture defines a mesh's appearance and details, e.g., scratches, patterns, colors, bumps, etc., which affect the realistic look of an image or an asset. A material determines the physical property of the surface, e.g., reflectivity, roughness, metallic, transmission, transparency, etc. To generalize the DL model and allow for effective TL, the texture and material properties need to be fine-tuned in order to match, as much as possible, the properties of real images [8]. Here, randomization can be used to allow for high-tolerance low-overfitting models which can detect a larger variation of the same assets through different sensors and under different environmental conditions (e.g., the same industrial asset viewed in different places of the factory floor and at different timeslots would look differently, depending on its position among its surrounding assets and environment, e.g., partly visible, positioned on a high-reflection surface, dimmed lighting, etc., cf. Figs. 8.1 and 8.2). Lighting consists of randomizing the physical light control parameters such as the light color, temperature, intensity, and directions. Additionally, the light and material components are interdependent and can impact each other. The properties of the material can cause a surface to reflect or absorb certain wavelengths of light, while the lighting conditions can change the asset's surface appearance of the material, resulting in various realistic and complex combinations [7].

Fig. 8.1 Structured domain randomization of three types of KLT load carriers from SORDI [7]

Fig. 8.2 Mixture of domain randomization components from a static camera viewport using SORDI assets [7]

8.1.3 Content Gap

The content gap represents the differences in the diversity, distribution, composition, placement, and behavior of assets or objects between the virtual environment and the real world [7]. In other words, the content gap is related to scene composition, i.e., how the virtual scene is similar to the real world in terms of the assets' physical positions, functionalities, and behaviors [9]. Generating synthetic images from a single scene with static assets might lead to model overfitting, without generalizing to the real-world. Here, data randomization provides a possible solution, transforming virtual scenes from static into dynamic environments where assets are generated, shown, or hidden at different positions of the 3D scene, following physical, logical, and semantic constraints inspired from the real world [6, 7]. For instance, Structured Domain Randomization (SDR) [9] is a type of domain randomization that takes into account the structure and context of the scene, according to the probability distributions of surrounding assets and environmental constraints. In this way, SDR-generated images allow the object recognition model to take the context around an object into consideration when performing the detection task (cf. Fig. 8.1). More specifically, the authors in [7] distinguish between (i) visibility randomization and (ii) transformation randomization. Visibility randomization follows a symmetric Bernoulli distribution for visualizing the asset or hiding it in the 3D space of the virtual scene. Transformation randomization represents an asset's transform settings consisting of position, rotation, and scale properties. It manipulates an assets' position along the x, y, and z axes, 3D rotates it around its pivot point, and changes its size in all dimensions respectively. Randomizing the position and rotation properties allows placing the object in a demarcated area at different orientations. Nevertheless, it is important to consider the objects' physical properties when spawning or replacing assets to avoid asset collisions or floating solids for example [7]. As a result, by combining different randomization components in a single scene, the same camera viewport is capable of rendering distinct and various images as shown in Fig. 8.2. We further describe domain randomization in more detail in Sect. 8.4.

8.2   Transfer Learning: A Promising Solution to the Reality Gap

A promising solution to address the reality gap is to perform Transfer Learning (TL), i.e., by transferring knowledge between domains or tasks. In a nutshell, TL is Machine Learning (ML) approach that aims at using knowledge gained from solving one task to solve another related task [10]. For instance, knowledge acquired while learning to recognize automobile parts can be used when attempting to recognize truck parts. In the latter example, we refer to the automobile parts as the source domain, and the truck parts as the target domain. Learning to recognize the source domain is conducted using typical ML (e.g., the domain of values consist of the automobile part images, associated with automobile part labels). Learning to recognize the target domain can also be performed using typical ML. Nonetheless, the main premise of TL suggests there is not enough data to train the target domain. Hence, the need to make use of the source domain learning in order to improve the target domain learning [11]. For instance, considering there are not enough labelled images of truck parts to perform ML on this target domain, TL provides a solution by using the models learned from the source domain, e.g., the automobile parts, in order to label truck parts. The source and target domains can share similar data (e.g., automobiles and trucks have similar shapes and sizes) and can share similar labels (e.g., automobile and truck parts belong to similar categories). TL aims to exploit the similarities between the source and target domains and labels, to help improve the learning of the target domain through the learning of the source domain [12].

In this context, simulation-to-real (sim-to-real) is a special case of TL where the source domain is a virtual simulation of the real world, i.e., synthetic data from the digital twin, while the target domain is the physical reality itself [6, 13]. Sim-to-real suggests the ML model be trained on synthetic data from the digital twin, given that amount of labeled synthetic data is enough to perform the learning task and achieve acceptable accuracy levels. In the context of BMW Group and Idealworks' use cases, synthetic SORDI images are utilized within the digital twin factory to train BMW's virtual robots, including the virtual iw.hub AMR. This allows omitting a huge amount of time from performing real image capture from the physical factory floor. Since the BMW digital twins and their SORDI assets are modelled in the image of their physical counterparts and closely resemble them, this means the learning performed in the digital twin can be easily transferred to the physical factory while maintaining acceptable quality levels. Nonetheless, if the virtual domain is different from the physical world, the model learned in the virtual domain will perform poorly when transferred to the real domain [10]. This phenomenon of performance loss from the virtual domain to the real domain is also referred to as the reality gap [13]. Two approaches can be used to reduce this gap [6]: (i) domain adaptation and (ii) domain randomization, which we describe in the following section.

8.3   Domain Adaptation

Domain adaptation aims at transforming the source domain into the target domain, or transforming both domains into a common domain, in order to boost the performance of TL and help reduce the reality gap [7, 14]. It allows a ML model trained with samples from a source domain to generalize to a target domain [14]. In the context of sim-to-real object recognition, it consists in generating photo-realistic images for the training dataset: the more the synthetic images resemble the real ones, the more the difference between the source (digital twin) and target (real world) domains is reduced, and therefore, the more the performance on real images is enhanced [6, 14]. An important amount of work has been achieved on domain adaptation, particularly for computer vision applications and more specifically for object recognition, e.g., [14–16]. Existing solutions can be roughly grouped in two main categories: (i) feature-level adaptation, and (ii) pixel-level adaptation. Feature-level domain adaptation focuses on learning domain-invariant features, either by learning a transformation of fixed or pre-computed features between the source and the target domains [17, 18], or by learning a domain-invariant feature extractor, usually represented by a Convolutional Neural Network (CNN) [19, 20]. Recent empirical evaluations have shown that the domain-invariant CNN-based approach usually provides improved results on a number of classification tasks, e.g., [21, 22]. Domain-invariance can be achieved by optimizing domain-level similarity metrics like maximum mean discrepancy [19], or by optimizing the response of an adversarially trained domain discriminator [22]. Pixel-level domain adaptation emphases on re-stylizing images from the source domain to make them look more like images from the target domain [23, 24]. Most methods in this category are based on image-conditioned Variational AutoEncoders (VAE) [25] or Generative Adversarial Networks (GANs) [26]. An original solution in [14] attempts to combine both feature-level and pixel-level domain adaptation for sim-to-real TL: given an initial set of synthetic images, the authors use a dedicated GAN model to produce adapted images that look more realistic. They consequently use the trained generator from the GAN model as a fixed module that adapts the synthetic visual input, while performing feature-level domain adaptation on extracted features that account for both the transferred images and the synthetic input. The authors in [14] focus on the robot grasping task as their application scenario, and show that, by using synthetic data and domain adaptation, their combined solution allows to reduce the number of real-world samples needed to achieve a given level of performance using only randomly generated simulated images.

8.4 Domain Randomization

Domain randomization [8, 27] is another approach to reduce the reality gap, by introducing variability and adding artificial noise to the synthetic training images. The main premise behind randomization is that the real world can be considered as just another random instance of the virtual simulation environment [6]. Achieving high variability in the synthetic images can be attained by using random camera and object positions, changing the lighting source and lighting conditions, and using non-realistic textures [7, 27]. The BMW Group TechOffice has recently adopted a domain randomization based approach into expand the usage of the SORDI dataset [7, 28], combining: (i) scene randomization, and (ii) camera randomization.

8.4.1 Scene Randomization

In order to scale the SORDI-based image generation pipeline to meet the need of industrial use cases, three levels of scene randomization are considered. The first level consists of creating a library of randomized composed assets, based on visibility randomization components to create various combinations between linked assets such as stacked KLTs as shown in Fig. 8.1, or different shapes, textures, and behaviors for the stillage assets, etc. Subsequently, the composed assets are placed in the scene by applying x and y position randomization and yaw rotation randomization within the scene's predefined subarea. The z-axis randomization is initially disregarded to consider the physical gravity parameter. This allows avoiding floating and colliding assets and preserving the scene content's realism [7]. The second level of randomization targets the walls, the ground, and the ceiling's materials and textures to simulate the background variability. The third level of randomization targets the light sources color, intensity, and rotation as shown in Fig. 8.2.

8.4.2   Camera Randomization

The camera is also a critical asset within the scene to display the virtual environment. Given its usage in a 3D environment, the camera system has 3D characteristics including visibility, position, and rotation. Accordingly, the camera movement and behavior in the scene affect the data capture. In this context, four main types of data capture are considered [7]: (i) Static Capture (SC), (ii) Full Capture Randomization (FCR), (iii) Constrained Capture Randomization (CCR), and iv) Sequential Capture (SeC). CS consists in fixing the camera at a static position and rotation. In this scenario, scene randomization is mandatory, otherwise, the same image is rendered at every frame. FCR consists in randomizing both the camera's source and the target points at any point in the scene's room. This allows diversifying the camera viewports and angle shots (e.g. high and low angles, tilted and point of view, long and close shots, aerial shots, etc.). On the one hand, randomizing the camera target only will be like a person standing in the same position looking around. On the other hand, randomizing the camera source only results in blind areas since it focuses on the same spot all the time. Therefore with FCS, the camera can be placed anywhere in the 3D scene, and can be looking at any point in the scene. CCR is similar to the FCR but omits camera capture from or at specific areas in complex and dense scenes. SeC stipulates that the camera is placed and follows a well-defined path to imitate the point of view of transportation robots (like the iw.hub). The TechOffice team uses NVIDIA GPUs to render visually realistic images out of the dense scenes produced for the BMW digital twins (cf. Fig. 8.2), where Isaac Sim cooperates with Omniverse to generate accurate annotations using 2D tight bounding boxes.

8.5   Real Image Obfuscation

Taking a step back from synthetic image generation and virtual environment randomization, we can safely state that all is not lost when it comes to creating larger real datasets! While using synthetic data is providing a viable solution to counter the lack of real industrial datasets, nonetheless, we can highlight a few promising directions to facilitate the creation of larger real datasets. Capturing real images from the factory floor might contain sensitive information such as individuals' faces, workers' belongings, or nametags. Due to privacy and security regulations, companies must guarantee a level of anonymization that prevents identifying the data subjects, by taking into account all the means likely to be used for identification [29]. In this context, several asset or region obfuscation techniques can be used to hide or remove sensitive information, including pixelation (also known as mosaicking) [30], blurring (Gaussian/motion) [31], and masking [32] (c.f. Fig. 8.3).

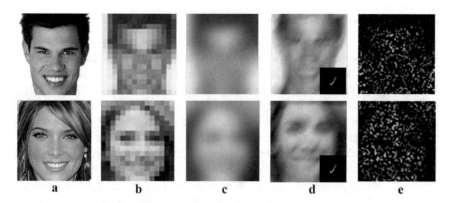

Fig. 8.3 Obfuscation techniques left to right: (**a**) Original clear image, (**b**) pixelated image (4×), (**c**) Gaussian blurred Image, (**d**) motion blurred and (**e**) masking by adding random black pixels [33]

8.5.1 Obfuscation Techniques

Obfuscation consists in altering or removing features from the images to hide sensitive information while keeping as much of the visual features as possible for the image to remain suitable for processing [33]. *Pixelating*, also referred to as mosaicking, is one of the earliest obfuscation techniques. The sensitive information to be obfuscated is divided into a square grid, i.e., a pixel box, where each pixel box is assigned one single color after averaging the values of the grouped pixels in it [30]. The size of the pixel box can be modified depending on the needed level of privacy. The larger the box, the more pixels will be averaged together, and the higher the level of privacy [33]. Though the size of the image stays the same, pixelating reduces the obfuscated region's resolution [34]. For instance, downscaling an image by a factor of four is equivalent to applying a pixel box of size 4 × 4 (c.f. Fig. 8.3b).

Blurring is another kind of degradation technique that can be utilized for obfuscation. It can be generated by a Gaussian kernel or via a camera motion effect, i.e., a motion blur. Gaussian like blur kernel is used extensively as an obfuscation technique [30], removing details from an image by applying a Gaussian kernel. A motion blur modifies the details of an image by generating the effect of a synthetic camera motion blur [31]. The level of blurriness is affected by the length and by the angle of the synthesized motion [33] (c.f. Fig. 8.3c, d).

Masking eliminates details from an image by replacing the original pixels by black pixels [32]. The masking technique can have numerous derivatives depending chiefly on the color intensity and location of the modified pixels. For example, if a person's face is considered sensitive, pixels can be modified around the eyes and mouth or at random points of the face [29]. The level of privacy depends on the amount, location, and color intensity of the altered pixels[33]. In Fig. 8.3e, random black pixels are considered around the entire face.

8.5.2   Lack of Privacy Guarantees

Existing obfuscation techniques in the context of images do not come with formal provable privacy guarantees [35]. To our knowledge, several evaluation frameworks have been proposed in the literature to experimentally evaluate the quality of obfuscation techniques applied on images and videos. Some frameworks rely on human observers [36] whereas others use quantitative metrics, namely structural similarity metrics [37] (e.g., Structural Similarity Index Measure – SSIM) [38], Peak Signal to Noise Ratio – PSNR [39], Structure Natural Measure – SNM [40], etc.). These evaluations usually consider (i) the efficiency of the privacy enhancement, (ii) the biometric utility preserved after privacy enhancement, or (iii) the robustness to attempts to reverse the obfuscation techniques [35]. Yet, despite the efforts to evaluate the quality of obfuscation techniques, evaluations remain empirical with no formal provable privacy guarantees. This may be related to the computational nature of obfuscation techniques. As described previously, obfuscation solutions attempt to change or remove certain features from the image to hide sensitive information. Nonetheless, the visual features that remain post-obfuscation can still be used to identify or reconstruct the obfuscated sensitive information using so-called obfuscation attacks [41]. The latter can be categorized as (i) recognition-based attacks and (ii) restoration-based attacks. *Recognition-based attacks*, e.g., [34, 42], rupture the images' privacy and anonymity by training learning-based algorithms to perform recognition tasks on obfuscated information. *Restoration-based attacks*, e.g., [43, 44], de-anonymize privacy-protected images by trying to restore and reconstruct the original features of the obfuscated information. Restoration and Recognition-based (R&R) attacks combine both techniques in order to recognize restored features of the obfuscated information [33, 34]. Several studies showed that DL-based solutions overtake traditional learning-based approaches for image restoration and recognition tasks, e.g., [45, 46]. Hence, from a privacy perspective, DL-based techniques are chosen as robust recognition-based and restoration-based attacks [47, 48].

8.5.3   Obfuscation Under Privacy Attacks

To address the privacy risks associated with obfuscation, a team of engineers from BMW Group and its research partners have designed a quantitative recommendation framework that evaluates the robustness of image obfuscation techniques and recommends the most resilient obfuscation solution against DL-assisted attacks [33]. In an initial attempt, the team assume that the background knowledge of the adversary comprises the obfuscation technique and its hyper-parameters, and suggest performing restoration-based attacks [33]. In a subsequent extension of the work, the team embeds and adapts a three-components adversary DL model inspired from [49] to perform facial image obfuscation (cf. Fig. 8.4). Several threat levels are

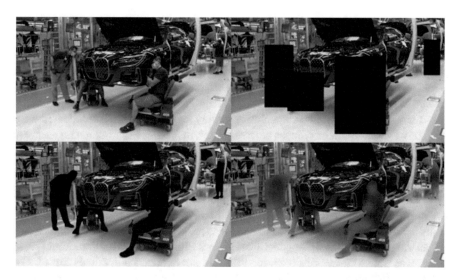

Fig. 8.4 Video snapshots from the BMW Group TechOffice Anonymization API that localizes and obfuscated sensitive information to preserve the BMW Group workers' identity on the factory floor

Table 8.1 Comparing the adversary's capabilities and knowledge with respect to the three threat levels [33]

Adversary's Components		Threat Levels	Level 1	Level 2	Level 3
Goal			Identify/Recover the identity of obfuscated faces		
Knowledge	External Knowledge	Public datasets	✓	✓	✓
	Background knowledge	Obfuscation technique	✓	✓	✓
		Obfuscation technique's hyper-parameters	✓	✓	✗
		Identities present in the target dataset	✗	✓	✓
Capabilities	DL-assisted Attacks	Restoration-based attack	✓	✓	✓
		Recognition-based attack	✗	✓	✓
		Restoration and Recognition-based attach	✗	✓	✓

defined with regard to the adversary's background knowledge which constitutes the obfuscation technique employed, its hyper-parameters, and the identities present in the target dataset. As stated previously, there is no a standardized evaluation methodology nor a defined model for adversaries when evaluating the robustness of image obfuscation [35], and more specifically face obfuscation techniques. Hence, several attacking scenarios are considered to explore new aspects of the adversary when evaluating the robustness of image face obfuscation, including restoration-based, recognition-based, and R&R-based attacks (cf. Table 8.1). The adversary's goal is to recover the identity of the obfuscated faces while its capabilities (i.e., restoration-based, recognition-based, or R&R-based attacks) depend heavily on its background knowledge (consisting of the obfuscation technique used and the identities present in the target dataset). Three threat levels are considered and mapped against the adversary's background knowledge, inspired by Shannon's Maxim.[1] Level 1 assumes the adversary is aware of the obfuscation technique used to obfuscate the target dataset along with its hyper-parameters. Level 2 assumes the adversary is aware of the identities present in the target dataset and of the obfuscation technique used along with its hyper-parameters. Level 3 assumes the adversary is aware of the identities present in the target dataset and of the obfuscation technique used but not of its hyper-parameters. A battery of experiments is conducted on a publicly available celebrity faces dataset [50]. A first set of experiments implements and evaluates the recommendation framework by considering four adversaries in Level 1 against four obfuscation techniques (e.g. pixelating, Gaussian blur, motion blur, and masking). A second set of experiments demonstrates how the adversary's attacking capabilities vary and scale with its knowledge in Level 2 and how it increases the potential risk of breaching the identities of blurred face images. A third set of experiments evaluates the possible privacy breaches and the attack range of an adversary in Level 3 against face images blurred with different kernels. The team was able to successfully re-identify 692 anonymized individuals out of 854 (81%) when simulating the strongest adversary, where the widest attack range was achieved when the training dataset of recognition-based attacks was prepared with a high-resolution blurring kernel of (37,37) [33].

8.5.4   Ongoing Directions

The TechOffice research team has been focused on processing people's faces because they are the most revealing in the context of images taken in industrial environments. Nonetheless, other visual features such as a person's name tag, posture, or personal belongings can be identifying and considered as sensitive information. Also, the present work limits the adversary's background knowledge to the identities present in the target dataset, allowing to mine images for each known identity and perform a DL-assisted attack to recognize and re-identify the identity of the obfuscated face images. Other scenarios can also be considered where the adversary's background knowledge can be limited to quasi-identifying information such as the individual's race or gender only. In that case, the adversary can perform DL-assisted attacks to recognize the gender or the race of the target individual [51] instead of recognizing the full identity of the person. This might lead to potential privacy breaches when linked to other data sources (e.g., identity disclosure via linking attacks [35]). Another ongoing direction is the investigation of different image classification solutions for identity recognition to trick, ruin, or corrupt DL models [33]. The team is investigating approaches that rely on designing adversarial examples by perturbing the query image at the inference phase either physically

---

[1] "The enemy knows the system", i.e., "one ought to design systems under the assumption that the enemy will immediately gain full familiarity with them".

(e.g. the target individual wears special accessories, e.g. glasses or hats [52]) or quantitatively (small perturbations are added on a pixel-level which are not visible to the human visual system [53]). The team is also experimenting with Intel's OpenVINO toolkit to accelerate and improve the model's efficiently when running the solution in online [54].

8.6 Facilitating Real Image Labelling

Another direction to facilitate the capture of larger real datasets concerns reducing the time needed by humans to annotate the captured images. After capturing the raw images and performing obfuscation to hide or remove the sensitive information, the images need to be annotated to highlight their contained assets. In order to perform annotation, the first step consists in locating the salient objects in the image, which is usually performed by superimposing bounding boxes over the objects' Region of Interest (RoI). The latter are subsequently annotated, and the labels are used to train the DL models for object recognition and other CV tasks. Here, multiple challenges arise. First, accurately labeling real image datasets requires excessive human effort and a considerable amount of time. Also, the object annotation process is highly prone to human error. Target objects might be wrongly labeled when a human annotator is biased to label one side of the target object. In addition, after the whole dataset has been labeled, the RoI of the target object might change because of industrial circumstances (e.g., certain objects change locations on the factory floor). In these cases, all labels need to be re-adjusted, resulting in unexpected delays and extra costs. In this context, many research efforts have focused at reducing the labeling time and avoiding crowdsourcing techniques by developing methods that are less time consuming than manual labeling. Most existing methods attempt to generate bounding-boxes around the target object, e.g., [55, 56]. Yet, to our knowledge, no attempts have been made in the literature to correct or refine inaccurate bounding boxes.

8.6.1 Bounding Box Automated Refinement

A team from BMW Group's TechOffice and Idealworks, in collaboration with research partners from academia, has addressed the bounding box refinement problem, aiming at enhancing the accuracy of existing bounding boxes. This allows obtaining better quality annotations to be used for training DL-based object recognition models, while reducing the time and effort for human intervention to manually correct the bounding boxes. For this purpose, the team has developed BAR (Bounding-box Automated Refinement), a Reinforcement Learning (RL) agent that learns from human examples to correct inaccurate annotations [57]. Instead of manually labeling new images to increase the dataset size, human effort is limited to correcting inaccurate annotations. After learning to find an optimal strategy to correct the bounding boxes based on their usage with the training images, BAR applies its knowledge on new images. Two approaches are considered to train BAR: an offline approach using RL in which the agent is trained by batches, and an online approach using Contextual Bandits (CB) in which the agent is re-trained after every new image. The advantages and limitations of both approaches have been assessed with different initializations, i.e. methods that generate initial bounding boxes. Results show for all initializations, at least one of the two approaches successfully improves the annotations, with an increase in Intersection-over-Union (IoU) with the ground-truths of up to 0.28, and a decrease in human intervention by 30–82%. BAR provides an online execution mode that improves some initializations with the advantage of real-time suggestions of new bounding boxes. BAR also provides an offline execution mode that produces more accurate bounding boxes and annotations, when tested with five different initializations, improving over state-of-the art solutions while reducing human effort.

8.6.2 Dynamics of Bounding Box Refinement

A practical example is shown in Fig. 8.5. Every image contains exactly one annotated target object whose bounding box is represented by its upper-left corner (*xmin, ymin*) and its lower-right corner (*xmax, ymax*). This bounding box is considered inaccurate if its IoU with the ground truth is below a certain threshold, denoted by β. Given an image and an inaccurate bounding box enclosing the target object, the goal of the agent is to correct the bounding box as shown in Fig. 8.5. The agent achieves this goal by executing a series of actions that modify the position and

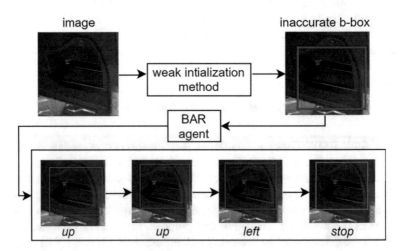

Fig. 8.5 BAR agent workflow during the testing phase (given an image and an inaccurate bounding box surrounding the target object, BAR chooses the path *[up,up,left]*) [57]

Table 8.2 Set of translation actions adopted in BAR [57]

Action	Corresponding equations	Action	Corresponding equations
Up	$xmin - c1 \times height$ $xmax - c1 \times height$	Wider	$ymin - c2 \times width$ $ymax + c2 \times width$
Down	$xmin + c1 \times height$ $xmax + c1 \times height$	Taller	$xmin - c2 \times height$ $xmax + c2 \times height$
Left	$ymin - c1 \times width$ $ymax - c1 \times width$	Fatter	$xmin + c2 \times height$ $xmax - c2 \times height$
Right	$ymin + c1 \times width$ $ymax + c1 \times width$	Thinner	$ymin + c2 \times width$ $ymax - c2 \times width$

aspect-ratio of the bounding box. This series of actions corresponds to an episode that ends with the final correction of the agent [57]. The dynamics of an episode are as follows [57]: At time step 0, the bounding box is given a weak initialization, i.e. a method that generates inaccurate bounding boxes. At time step 1, the agent performs an action and determines the quality of the new bounding box based on its IoU with the ground-truth. If the IoU is below β, the agent needs to modify the bounding box in the next time step; otherwise, the agent stops and the episode terminates. The length of an episode is limited in number of actions to prevent cases where the agent fails to converge to an accurate bounding box. During an episode, the agent is allowed to either move according to a set of predefined translation actions listed in Table 8.2, or to end the episode by choosing the *stop* action. Table 8.2 reports the equations according to which each of the translation actions changes the corners of the bounding box, where $c1$ is the percentile of the box's current dimensions (width or height) that is added to or removed from its coordinates. Given the goal of the agent and the dynamics of an episode, the bounding box correction problem can be formulated as a Sequential Decision Making Problem (SDMP), specifically, an episodic RL problem [57]. The agent interacts with the environment to learn a policy that determines the optimal path to change an initial inaccurate bounding box into an accurate one. The path is composed of a list of translation actions T, and a terminal action E that ends the correction. The task of the agent for the path TE is therefore to find: (i) the optimal actions T that modify the bounding box, and (ii) the step at which to end the episode when an accurate b-box is reached [57].

8.6.3 BMW LabelTool Lite

The general process of image annotation starts by locating the salient objects in the image using bounding boxes, and then associating labels with each bounding box describing the semantics of its contained object. In this context, the team at BMW Group and Idealworks has developed the *BMW LabelTool lite*, an easy to use image data annotation tool with little to no configuration needed [58]. While BAR helps adjust the bounding boxes to prepare for annotation, *LabelTool lite* completes BAR

Fig. 8.6 Snapshot of the BMW LabelTool lite interface

by helping reduce the effort and time needed by humans to label the bounding boxes (cf. Fig. 8.6). The lite version of the tool allows users to easily deploy the solution and start labeling their images for state-of-the-art DL training purposes with a dockerized implementation solution. Users can also directly use the labels provided by the *LabelTool lite* to train with the TechOffice's Yolo and Tensorflow Training GUI repositories [59]. Moreover, it is possible to connect a pre-trained or a custom-trained model to the *LabelTool lite*. This functionality allows accelerating the labeling process whereby the connected model can be actively used to suggest appropriate labels for each image. The tool offers different options that allow comfortable navigation through the data set while labeling (e.g., navigate to the next image in the data set – ">", navigate to the previous image in the data set – "<", navigate to the next image that has no bounding boxes – ">>", navigate to a particular image by inputting the image number directly, and navigate to any image by clicking or dragging the cursor on the scroll line).

In addition to the above mentioned features, the *LabelTool lite* offers zoom-in and zoom-out functionality for the images, increase and decrease of brightness for the images, new images can be uploaded to the data set, image attributes (name and resolution) can be displayed for each image, and images can be deleted (one at a time) along with the corresponding bounding boxes. The tool also offers a variety of functionalities for bounding box creation including resizing the boxes after creation, moving the boxes via simple drag and drop functionality, copying the boxes and pasting them as needed, setting them to *fill* or *unfill* as needed, and controlling the line thickness of the bounding boxes according to the user's preferences.

8.6.4   Ongoing Directions

The current BAR framework is designed to process single object images, and can be applicable to multiple objects per image as long as they do not overlap. An ongoing direction focuses on extending the current framework to handle overlapping objects. Since human intervention through correction is an important factor in training, this task can be formulated in an active learning setting in which the agent asks for human feedback while training when it is uncertain [57]. This can also help in reducing the effect of noisy data provided by human annotators. Another direction is to introduce a dynamic version of BAR which chooses to use either Reinforcement Learning (RL) or Contextual Bandits (CB) methods depending on the use-case and image structural similarity of the dataset. Furthermore, the team is investigating real-life measurements on a sample population of human annotators to obtain a better measure of the human intervention being saved. Using a continuous rather than a fixed parameter for translation actions would also allow BAR to better adapt to different types of initializations independently from parameter tuning [57].

8.7   Mixing Real and Synthetic Datasets

Going back to the main motivation behind the creation of SORDI, most existing real image industrial datasets are small and insufficient for CV and object recognition training. Hence the need to supplement the real data with synthetic data while guaranteeing a minimal reality gap between both real and synthetic domains. In this context, multiple solutions are being investigated to mix real and synthetic datasets in order to train their object recognition models [60]. The team aims at better understanding how mixing multi-domain datasets can affect the object recognition task. Experiments are currently concentrated on the load carrier (KLT) box (cf. Fig. 8.7),

a   b

Fig. 8.7 Real and synthetic KLT box image assets. (**a**). Real KLT box. (**b**). Synthetic KLT box from SORDI

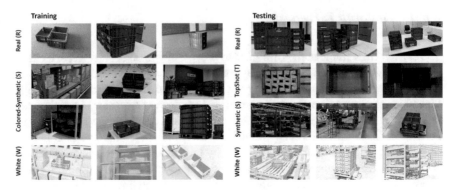

Fig. 8.8 Different domain KLT box training and testing datasets [60]

a modular and ergonomic industrial asset that is easy to use and store. Despite its effectiveness, this asset is hard to detect accurately since it appears in intricate variations on the factory floor (cf. Fig. 8.8). The team is gradually mixing datasets from several domains, acquiring real-captured images using cameras deployed on the factory floor, and generating the corresponding synthetic images from the TechOffice's digital twin implemented using NVIDIA's Omniverse. The team is studying the efficient mix of images from different domains including a source domain obtained from any related dataset, and a target domain related to the robot's application domain with the purpose of training the object recognition model accordingly. Experiments are performed with different combinations and ratios of source and destination domains (real images R, colored synthetic images S, and white synthetic images W) and evaluated on multiple datasets (both real and synthetic domains) as shown in Fig. 8.9.

8.7.1   Real Data Acquisition

High-resolution (1080p) videos are recorded at a rate of 30 frames per second (fps), while moving around KLT boxes that are placed in different rooms, on different surfaces, and in multiple light conditions (e.g., perfect indoor lighting, curtain shadows, backlight, outdoor lighting only, etc.). Every video contains low-angle and top-shot viewports taken from far to near distances of the KLTs. The best frames are extracted from the videos, by dividing the video into equal batches, and selecting from each batch the sharpest frame with the lowest Laplacian filter metric [60]. Instead of hand annotating all real images from scratch, the annotations are inferred using existing pre-trained KLT object recognition models and the predicted bounding boxes are fine-tuned automatically as described in Sect. 8.6. Images are considered for each training dataset divided as follows: for each ratio r of the source or domain, $(1 - r)$ is considered for the target domain data where r and $(1 - r)$ ratios of images are uniformly selected from both domains respectively. Afterward, the domains are mixed and shuffled to compose a single hybrid dataset [60].

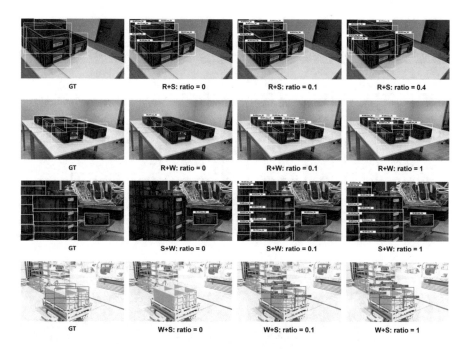

Fig. 8.9 Bounding box predictions using R + S, R + W, and S + W models at different mix ratio [60]

8.7.2 Model Training

Initial experiments are performed with a total of 7, 500 training images, gradually combining two domains images by a step ratio equal to 0.1, from 0 to 1, where 0 indicates a dataset of 7500 images from the target domain while 1 refers to a 100% images from the source domain. Multiple TL models are considered, including TensorFlow2's FRCNN Resnet-50, FRCNN Resnet-101, SSD MobileNet v2, FPNLite, and SSD efficientNet d1 detection models pre-trained on the Microsoft COCO dataset [2]. Multiple hybrid datasets are considered, including real and plain color synthetic (R + S), real and white color synthetic (R + W), and plain color and white color synthetic (S + W, cf. Fig. 8.9). The trained models are evaluated on both (i) real images: 850 images with different distributions (R), including top shot images (T) from different cameras, and (ii) synthetic images: 850 colored images (S) and 850 white-shaded images (W) in the digital twin virtual environment. Considering the KLT object recognition problem, Average Precision (AP) results summarized in Table 8.3 show that hybrid datasets comprising 20–30% of images similar to the test domain allow achieving nearly maximum detection accuracy with AP > 90%.

This work is currently being extended to support other challenging logistic assets (e.g., stillage, dolly, jack, and AMR). Also, new randomizations are currently being developed, including textures, occlusions, light and camera parameters, postprocessing filters, and augmentations, etc. New directions including image compressions (e.g., lossy and lossless), and data augmentations (e.g., context and

Table 8.3 AP@0.7 gradual evaluation of FRCNN Resnet-50 models trained on mixed domain datasets [60]

Ratio	R + S:R	R + W:R	R + S:T	R + W:T	S + R:S	S + W:S	R + W:W	S + W:W	R + S:R + S
0.00	84.77	18.29	73.24	0.00	5.49	0.59	**45.07**	**44.24**	71.30
0.10	88.68	89.61	87.29	43.97	45.33	53.93	42.49	44.17	73.61
0.20	89.24	89.61	82.75	62.48	50.59	55.82	42.74	44.18	**74.20**
0.30	89.69	89.89	87.07	59.03	52.80	55.87	41.91	43.67	73.29
0.40	89.84	90.71	80.11	71.13	55.14	56.94	40.69	43.04	73.51
0.50	90.78	90.71	87.88	61.83	54.41	56.64	39.69	42.95	72.20
0.60	**90.88**	90.75	86.98	72.96	56.81	57.76	38.66	41.96	72.58
0.70	90.71	90.70	**91.02**	73.37	57.25	57.01	37.73	41.99	70.87
0.80	90.87	90.45	88.23	65.00	**58.60**	58.19	34.91	40.44	70.25
0.90	90.40	90.54	84.23	71.07	58.33	58.04	31.52	37.44	66.79
1.00	90.74	**90.93**	78.42	**75.53**	58.14	**58.85**	1.37	0.00	54.68

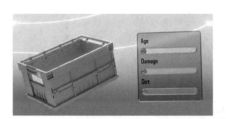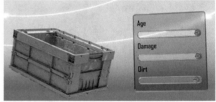

a

b

Fig. 8.10 Different material degradation and context augmentation simulations using SORDI. (**a**). Material degradation parameter variation. (**b**). Context augmentation and variation of KLT asse

neighborhood), material degradation (e.g., age and degradation), asset dynamicity (according to physical properties, e.g., weight and surface drag) and different representation modalities (e.g., color and texture) are currently being investigated to optimize the hybrid training pipeline and minimize the dataset size while maintaining good quality levels (cf. Fig. 8.10).

8.8 Toward Green Manufacturing

Unleashing the power of synthetic assets through SORDI! Promoting the usage of AI and CV through synthetic and hybrid datasets! Promoting AI-enabled robotics automation on the factory floor! Helping develop BMW Group's next generation immersive digital factories! The latter summarize some of the latest technological breakthroughs that are pioneered by BMW Group's TechOffice, NVIDIA, Microsoft, and Idealworks, and their industrial and academic partners. But there is more to this success story… Digitalization is at the center of BMW Group's iFactory vision [61]: a revolutionary strategy for the automotive production of the future… The TechOffice and Iealworks' technological breakthroughs are right smack in the middle of the iFactory vision! In promoting the BMW iFactory model, BMW Group is redefining the future orientation of its plants and is putting forth new standards in climate protection and competitiveness with flexible, efficient, sustainable and digital manufacturing technologies [62]. The iFactory vision is focused on three main areas [62]: (i) Lean, (ii) Green, and (iii) Digital. Lean production stands for efficient and highly flexible production, which makes processes more capable of integration and more variable. This will result in greater agility, leaner processes, and more-competitive structures. Green production stands for using the latest technologies to promote resource-saving and circular industry. Digital production stands for using the latest digitalization technologies from virtualization, AI, and data processing to digitalize the automotive supply chain and production pipelines. In this context, we view digital production as the main pillar that will help realize its lean and green counterparts. Milan Nedeljković, member of the Board of Management for Production at BMW Group, expressed it eloquently: "The BMW iFactory delivers – on our desire, as a member of society, to support climate protection and sustainability. We are using digitalization to make this happen – while remaining absolutely competitive" [61]. In fact, BMW Group has set itself a firm goal for 2030 throughout its entire production chain: to reduce CO2 emissions by at least 40%, starting from acquiring the raw material, to the supply chain, production and use phases, all the way to recycling [63]. This is a central step, in line with the ambitious objective of the Paris Climate Agreement that commits BMW Group to the goal of total climate neutrality by latest 2050 [63]. Digitalization and sustainability are complementary and intertwined steps that are central toward achieving a full-fledged circular economy, according to BMW Group's corporate strategies [64], especially when digital innovations are developed into effective use cases for production [61]. Whether it be designing new tools and machines, creating new industrial processes to optimize production, or inventing new processes focused on alternative energy sources, the push toward full-fledged factory digitalization through immersive digital twins and synthetic assets would allow engineers and designers to imagine new solutions, implement them in the virtual factory, test them on the digital twin's virtual factory floor, and simulate their impact and sustainability before ever deploying them in the physical plant. "With digitalization, we are achieving a new dimension of data consistency throughout the value chain and across all process chains… Data science, AI, and virtualization are making the BMW iFactory digital. We are at home in the digital world – and the digital world is at home in the iFactory" said Nedeljković in his 2022 interview about the iFactory [61]. This is equally emphasized by Oliver Zipse, BMW Group's Chairman of the Board of Management who sums it up powerfully [64]: "Our digitalization push comes at exactly the right time… We are gearing our entire production networks toward e-mobility, and the NEUE KLASSE. Our plants are becoming the BMW iFactory".

References

1. O. Russakovsky et al., ImageNet large scale visual recognition challenge. Int. J. Comput. Vis. **115**(3), 211–252 (2015)
2. T. Lin et al., *Microsoft coco: Common Objects in Context.* European Conference on Computer Vision (Springer, 2014). pp. 740–755
3. S. Ren et al., Faster R-CNN: Towards real-time object detection with region proposal networks. Adv. Neural Inf. Proces. Syst. **28**, 91–99 (2015)
4. W. Liu et al., *SSD: Single shot MultiBox detector.* European Conference on Computer Vision (ECCV), 2016. pp. 21–37
5. J. Terven and D. Esparza, *A Comprehensive Review of YOLO: From YOLOv1 to YOLOv8 and Beyond.* CoRR abs/2304.00501, 2023
6. D. Horváth et al., Object detection using Sim2Real domain randomization for robotic applications. IEEE Trans. Rob. **39**(2), 1225–1243 (2023)
7. Akar C. Abou et al., *SORDI.ai: Large-Scale Synthetic Object Recognition Dataset Generation for Industries* (Technical Report, BMW Group TechOffice, 2023). https://sordi.ai/research/sordi-2023
8. L. Eversberg, J. Lambrecht, Generating images with physics-based rendering for an industrial object detection task: Realism versus domain randomization. Sensors **21**(23), 7901 (2021)
9. A. Prakash et al., *Structured Domain Randomization: Bridging the Reality Gap by Context-Aware Synthetic Data.* International Conference on Robotics and Automation (ICRA'19), 2019. pp. 7249–7255
10. K. Weiss et al., A survey of transfer learning. J. Big Data **3**(1), 9 (2016). https://doi.org/10.1186/s40537-016-0043-6
11. S. Yao et al., A survey of transfer learning for machinery diagnostics and prognostics. Artif. Intell. Rev. **56**(4), 2871–2922 (2023)
12. S. Dai, F. Meng, Addressing modern and practical challenges in machine learning: A survey of online federated and transfer learning. Appl. Intell. **53**(9), 11045–11072 (2023)
13. A. Pashevich, et al., *Learning to Augment Synthetic Images for Sim2Real Policy Transfer.* IEEE/RJS International Conference on Intelligent RObots and Systems (IROS'19), 2019. pp. 2651–2657
14. K. Bousmalis et al., *Using Simulation and Domain Adaptation to Improve Efficiency of Deep Robotic Grasping.* IEEE International Conference on Robotics and Automation (ICRA'18), 2018. pp. 4243–4250
15. G. Csurka, A comprehensive survey on domain adaptation for visual applications, in *Domain Adaptation in Computer Vision Applications* (2017). pp. 1–35
16. V. Patel et al., Visual domain adaptation: A survey of recent advances. IEEE Signal Process. Mag. **32**(3), 53–69 (2015)
17. R. Caseiro et al., *Beyond the Shortest Path: Unsupervised Domain Adaptation by Sampling Subspaces Along the Spline Flow.* Computer Vision and Pattern Recognition (CVP'15), 2015. pp. 3846–3854
18. B. Sun et al., *Return of Frustratingly Easy Domain Adaptation.* AAAI Conference on Artificial Intelligence (AAAI'16), 2016. pp. 2058–2065
19. K. Bousmalis et al., *Domain Separation Networks.* Conference on Neural Information Processing Systems (NeurIPS'16), 2016. pp. 343–351

20. Y. Ganin et al., Domain-adversarial training of neural networks. J. Mach. Learn. Res. **17**, 2096 (2016)

21. Y. Di et al., *Fault Diagnosis of Rotating Machinery Based on Domain Adversarial Training of Neural Networks*. IEEE International Symposium on Industrial Electronics (ISIE'21), 2021. pp. 1–6

22. A. Gallego et al., Incremental unsupervised domain-adversarial training of neural networks. IEEE Trans. Neural Netw. Learn. Syst. **32**(11), 4864–4878 (2021)

23. K. Bousmalis et al., *Unsupervised Pixel-Level Domain Adaptation with Generative Adversarial Networks*. Computer Vision and Pattern Recognition (CVPR'17), 2017. pp. 95–104

24. C. Xu et al., Unsupervised domain adaption with pixel-level discriminator for image-aware layout generation. CoRR abs/2303.14377, 2023

25. H. Yang et al., Synthesizing multi-contrast MR images via novel 3D conditional variational auto-encoding GAN. Mobile Netw. Appl. **26**(1), 415–424 (2021)

26. H. Tang et al., AttentionGAN: Unpaired image-to-image translation using attention-guided generative adversarial networks. IEEE Trans. Neural Netw. Learn. Syst. **34**(4), 1972–1987 (2023)

27. J. Tobin et al., *Domain Randomization for Transferring Deep Neural Networks from Simulation to the Real World*. IEEE/RSJ International Conference on Intelligent Robots and Systems (IROS'17), 2017. pp. 23–30

28. Akar C. Abou et al., *Synthetic Object Recognition Dataset for Industries*. International Conference on Graphics, Patterns and Images (SIBGRAPI'22), 2022. pp. 150–155

29. J. Tekli et al., (k, l)-Clustering for Transactional Data Streams Anonymization, in *Information Security Practice and Experience* (2018). pp. 544–556

30. S. Hill et al., *On the (In)effectiveness of Mosaicing and Blurring as Tools for Document Redaction*. Proceedings on Privacy Enhancing Technologies (PoPETs'16), 2016. pp. 403–417

31. G. Boracchi, A. Foi, Modeling the performance of image restoration from motion blur. IEEE Trans. Image Process. **21**(8), 3502–3517 (2012)

32. R. Fellin, M. Ceccato, Experimental assessment of XOR-masking data obfuscation based on K-clique opaque constants. J. Syst. Softw. **162** (2020)

33. J. Tekli, B. al Bouna et al., *A Framework for Evaluating Image Obfuscation under Deep Learning-Assisted Privacy Attacks*. Conference on Privacy, Security and Trust (PST'19), 2019. pp. 1–10

34. R. McPherson et al., Defeating image obfuscation with deep learning. CoRR abs/1609.00408, 2016

35. B. Meden et al., Privacy-enhancing face biometrics: A comprehensive survey. IEEE Trans. Inf. Forensics Secur. **16**, 4147–4183 (2021)

36. Y. Li et al., Effectiveness and users' experience of obfuscation as a privacy-enhancing technology for sharing photos. Proc. ACM Hum.-Comput. Interact. **67**(1–67), 24 (2017)

37. T. Nawaz et al., Effective evaluation of privacy protection techniques in visible and thermal imagery. J. Electron. Imaging **26**(5), 51408 (2017)

38. Z. Wang et al., Image quality assessment: From error measurement to structural similarity. IEEE Trans. Image Process. **13**, 4 (2004)

39. S.R. Al, J. Tekli, Comparing deep learning models for low-light natural scene image enhancement and their impact on object detection and cCassification: Overview, empirical evaluation, and challenges. Signal Process. Image Commun. **109**, 116848 (2022)

40. Sobbahi R. Al and J. Tekli, *Low-Light Homomorphic Filtering Network for Integrating Image Enhancement and Classification*. Signal Processing – Image Communication (SPIC), 2022. pp. 100: 116527

41. A. Redondo and D. Insua, *Protecting from Malware Obfuscation Attacks Through Adversarial Risk Analysis*. CoRR abs/1911.03653, 2019

42. E. Newton et al., Preserving privacy by De-identifying face images. IEEE Trans. Knowl. Data Eng. **17**(2), 232–243 (2005)

43. D. Abramian and A. Eklund, *Refacing: Reconstructing Anonymized Facial Features Using GANs*. IEEE International Symposium on Biomedical Imaging (ISBI'19), 2019. pp. 1104–1108

44. N. Ruchaud and J. Dugelay, *Automatic Face Anonymization in Visual Data: Are We Really Well Protected?* Image Processing: Algorithms and Systems (IPAS'16), 2016. pp. 1–7

45. S. Dargan et al., A survey of deep learning and its applications: A new paradigm to machine learning. Arch. Comput. Methods Eng. **17**, 1071–1092

46. W. Yang et al., Deep learning for single image super-resolution: A brief review. IEEE Trans. Multimed. **21**(12), 3106–3121 (2019)

47. H. Hao et al., *Robustness Analysis of Face Obscuration.* IEEE International Conference on Automatic Face & Gesture Recognition (FG'20), 2020. pp. 176–183

48. K. Packhauser et al., Is medical chest X-ray data anonymous? CoRR abs/2103.08562, 2021

49. Q. Do et al., The role of the adversary model in applied security research. Comput. Secur. **81**, 156–181 (2019)

50. Z. Liu et al., *Deep Learning Face Attributes in the Wild.* In Proceedings of International Conference on Computer Vision (ICCV'15), 2015. pp. 3730–3738

51. Y. Linwei , et al., *Privacy-Preserving Age Estimation for Content Rating.* IEEE International Workshop on Multimedia Signal Processing (MMSP'18), 2018. pp. 1–6

52. S. Komkov and A. Petiushko, *AdvHat: Real-World Adversarial Attack on ArcFace Face ID System.* International Conference on Pattern Recognition (ICPR'20), 2020. pp. 819–826

53. I. Goodfellow et al., *Explaining and Harnessing Adversarial Examples.* International Conference on Learning Representations (ICLR'15) (Posters, 2015)

54. Intel, *AI-based Quality Control on Evern PC for Every Employee: BMA Group is Bankin on Intel OpenVINO* (Solution Brief, 2021). https://www.robotron.de/fileadmin/Robotron_DE/Dokumente/Sontiges/2021-06-intel-solution-brief-bmw-robotron-en-final.pdf

55. D. Papadopoulos et al., *We Don't Need No Bounding-Boxes: Training Object Class Detectors Using Only Human Verification.* Computer Vision and Pattern Recognition (CVPR'16), 2016. pp. 854–863

56. D. Papadopoulos, et al., *Training Object Class Detectors with Click Supervision.* Computer Vision and Pattern Recognition (CVPR'17), 2017. pp. 180–189

57. M. Ayle et al., Bar – A reinforcement learning agent for bounding-box automated refinement. Proc. AAAI Conf. Artif. Intell. **34**(03), 2561–2568 (2020)

58. BMW Innovation Lab, *BMW Labeltool Lite,* 2023. https://github.com/BMW-InnovationLab/BMW-Labeltool-Lite

59. BMW Innovation Lab, *BMW YOLOv4 Training Automation,* 2023. https://github.com/BMW-InnovationLab/BMW-YOLOv4-Training-Automation

60. Akar C. Abou et al., *Mixing Domains For Smartly Picking and Using Limited Datasets in Industrial Object Detection* (Technical Report, BMW Group TechOffice, 2023). https://www.sordi.ai/research/mixing-domains-for-smartly-picking-and-using-limited-datasets-in-industrial-object-detection. Accepted for presentation and publication in the proc. of the Inter. Conf. on Computer Vision Systems (ICVS 2023)

61. BMW Group, *We Are All Designers of the BMW iFactory* (BMW Group News, 2022). https://www.bmwgroup.com/en/news/general/2022/interview-bmw-ifactory.html

62. BMW Group, *Production of Tomorrow: BMW iFactory* (BMW Group News, 2022). https://www.bmwgroup.com/en/news/general/2022/bmw-ifactory.html

63. BMW Group, *We Make the BMW Group Sustainable* (BMW Group, 2023). https://www.bmw-group.com/en/sustainability.html

64. BMW Group, *BMW Group Report 2022,* 2022. https://www.bmwgroup.com/content/dam/grpw/websites/bmwgroup_com/ir/downloads/en/2023/bericht/BMW-Group-Report-2022-en.pdf

Printed in the United States
by Baker & Taylor Publisher Services